ASCE Manuals and Reports on Engineering Practice No. 81

Guidelines for Cloud Seeding to Augment Precipitation

Second Edition

Submitted by
the Standards Committee on Atmospheric Water Management (AWM) to the Standards Development Council (SDC) of the Environmental and Water Resources Institute (EWRI) of the American Society of Civil Engineers (ASCE)

Completed by
Revision of ASCE Manual No. 81 Subcommittee Editors: Conrad G. Keyes, Jr. (Chief Editor); Co-Editors: Bruce A. Boe, George W. Bomar, Robert R. Czys, Thomas P. DeFelice, and Don A. Griffith

ASCE Published by the American Society of Civil Engineers

Environmental & Water Resources Institute of the American Society of Civil Engineers EWRI

Library of Congress Cataloging-in-Publication Data

Guidelines for cloud seeding to augment precipitation.—2nd ed. / completed by . . . subcommittee editors, Conrad G. Keyes, Jr., . . . [et al.].
p. cm. —(ASCE manuals and reports on engineering practice; no. 81)
Includes bibliographical references.
ISBN 0-7844-0819-X
1. Rain-making—United States—Handbooks, manuals, etc. 2. Precipitation (Meteorology)—Modification—United States—Handbooks, manuals, etc. I. Keyes, Conrad G. II. American Society of Civil Engineers. III. Series.

QC928.7.G85 2006
551.68'76—dc22

2005029297

Published by American Society of Civil Engineers
1801 Alexander Bell Drive
Reston, Virginia 20191
www.pubs.asce.org

ISBN 07844-0819-X
Manufactured in the United States of America.

MANUALS AND REPORTS ON ENGINEERING PRACTICE

(As developed by the ASCE Technical Procedures Committee, July 1930, and revised March 1935, February 1962, and April 1982)

A manual or report in this series consists of an orderly presentation of facts on a particular subject, supplemented by an analysis of limitations and applications of these facts. It contains information useful to the average engineer in his everyday work, rather than the findings that may be useful only occasionally or rarely. It is not in any sense a "standard," however; nor is it so elementary or so conclusive as to provide a "rule of thumb" for nonengineers.

Furthermore, material in this series, in distinction from a paper (which expressed only one person's observations or opinions), is the work of a committee or group selected to assemble and express information on a specific topic. As often as practicable the committee is under the direction of one or more of the Technical Divisions and Councils, and the product evolved has been subjected to review by the Executive Committee of the Division or Council. As a step in the process of this review, proposed manuscripts are often brought before the members of the Technical Divisions and Councils for comment, which may serve as the basis for improvement. When published, each work shows the names of the committees by which it was compiled and indicates clearly the several processes through which it has passed in review, in order that its merit may be definitely understood.

In February 1962 (and revised in April 1982) the Board of Direction voted to establish:

A series entitled "Manuals and Reports on Engineering Practice," to include the Manuals published and authorized to date, future Manuals of Professional Practice, and Reports on Engineering Practice. All such Manual or Report material of the Society would have been refereed in a manner approved by the Board Committee on Publications and would be bound, with applicable discussion, in books similar to past Manuals. Numbering would be consecutive and would be a continuation of present Manual numbers. In some cases of reports of joint committees, bypassing of Journal publications may be authorized.

MANUALS AND REPORTS ON ENGINEERING PRACTICE

No.	Title
13	Filtering Materials for Sewage Treatment Plants
14	Accommodation of Utility Plant Within the Rights-of-Way of Urban Streets and Highways
35	A List of Translations of Foreign Literature on Hydraulics
40	Ground Water Management
41	Plastic Design in Steel: A Guide and Commentary
45	Consulting Engineering: A Guide for the Engagement of Engineering Services
46	Pipeline Route Selection for Rural and Cross-Country Pipelines
47	Selected Abstracts on Structural Applications of Plastics
49	Urban Planning Guide
50	Planning and Design Guidelines for Small Craft Harbors
51	Survey of Current Structural Research
52	Guide for the Design of Steel Transmission Towers
53	Criteria for Maintenance of Multilane Highways
54	Sedimentation Engineering
55	Guide to Employment Conditions for Civil Engineers
57	Management, Operation and Maintenance of Irrigation and Drainage Systems
59	Computer Pricing Practices
60	Gravity Sanitary Sewer Design and Construction
62	Existing Sewer Evaluation and Rehabilitation
63	Structural Plastics Design Manual
64	Manual on Engineering Surveying
65	Construction Cost Control
66	Structural Plastics Selection Manual
67	Wind Tunnel Studies of Buildings and Structures
68	Aeration: A Wastewater Treatment Process
69	Sulfide in Wastewater Collection and Treatment Systems
70	Evapotranspiration and Irrigation Water Requirements
71	Agricultural Salinity Assessment and Management
72	Design of Steel Transmission Pole Structures
73	Quality in the Constructed Project: A Guide for Owners, Designers, and Constructors
74	Guidelines for Electrical Transmission Line Structural Loading
76	Design of Municipal Wastewater Treatment Plants
77	Design and Construction of Urban Stormwater Management Systems
78	Structural Fire Protection
79	Steel Penstocks
80	Ship Channel Design
81	Guidelines for Cloud Seeding to Augment Precipitation
82	Odor Control in Wastewater Treatment Plants
83	Environmental Site Investigation
84	Mechanical Connections in Wood Structures
85	Quality of Ground Water
86	Operation and Maintenance of Ground Water Facilities
87	Urban Runoff Quality Manual
88	Management of Water Treatment Plant Residuals
89	Pipeline Crossings
90	Guide to Structural Optimization
91	Design of Guyed Electrical Transmission Structures
92	Manhole Inspection and Rehabilitation
93	Crane Safety on Construction Sites
94	Inland Navigation: Locks, Dams, and Channels
95	Urban Subsurface Drainage
96	Guide to Improved Earthquake Performance of Electric Power Systems
97	Hydraulic Modeling: Concepts and Practice
98	Conveyance of Residuals from Water and Wastewater Treatment
99	Environmental Site Characterization and Remediation Design Guidance
100	Groundwater Contamination by Organic Pollutants: Analysis and Remediation
101	Underwater Investigations
103	Guide to Hiring and Retaining Great Civil Engineers
104	Recommended Practice for Fiber-Reinforced Polymer Products for Overhead Utility Line Structures
105	Animal Waste Containment in Lagoons
106	Horizontal Auger Boring Projects
107	Ship Channel Design and Operation
108	Pipeline Design for Installation by Horizontal Directional Drilling
109	Biologic Nutrient Removal Operation in Wastewater Treatment Plants

CONTENTS

TABLES

FIGURES

FOREWORD

By Conrad G. Keyes, Jr., ScD, P.E., P.S., D.WRE, WMA CM/CO, Hon. M. ASCE (Life), F. NSPE (Life)

Traditional water resources management pertains to making reasonable use of available water, desalinization, and minimizing loss due to floods. Atmospheric water management provides a cost-effective means for augmenting available water and reducing damage during meteorological events.

In many areas of the United States and the world, there is a need for new water supplies. These updated guidelines are intended to provide water resources managers and others with information and references they will need for decision making regarding the use of cloud seeding to augment available water supplies.

The Manual incorporates pertinent background on the science and practice of weather modification by cloud seeding to augment precipitation. Legal, social, environmental, and economic factors motivating and limiting operational cloud seeding are reviewed. The technologies, instrumentation and procedures needed to implement a cloud seeding program are described. This is all intended to give the water resources manager the broad spectrum and the practical details of what is involved in utilizing the cloud seeding (atmospheric water management) technology.

The American Society of Civil Engineers (ASCE) Weather Modification Committee (1960–1985) and the Climate and Weather Change Committee (1985–1996) were fortunate enough to bring together experts in the weather modification field and have them devote a great amount of volunteer time to write the first versions of this most valuable document. The 1982 Weather Modification Committee, the 1993 Climate and Weather Change Committee, and the 1982 and 1994 Executive Committees of the Irrigation and Drainage Division are to be commended for their thorough and helpful review of the first document that was published in the ASCE Journal of Irrigation and Drainage Engineering in March 1983, pp. 111–182 (parts written by Paul C. Summers – Foreword, Robert D. Elliott – Summary, Olin H. Foehner – SEE Issues, Ray Jay Davis – Legal Aspects, Lewis O. Grant – Scientific Basis, Don A. Griffith – Modes & Instrumentation, and Conrad G. Keyes, Jr. – How to Implement.)

The original Task Committee appreciated the extensive technical editing of each section of the manual by the personnel of OPHIR Corporation. The Consortium of Atmospheric Resources Development provided funds for the review of the first version of the Manual, and the North American Interstate Weather Modification Council provided funds for travel to a meeting of the 1992–1993 Task Committee involved in the revision of the 1983 Guidelines published by ASCE.

The 1995 Manual was authored by the following individuals (by section): (1) Robert D. Elliott, Conrad G. Keyes, Jr., and Roger F. Reinking; (2) Roger F. Reinking, Neil H. Berg, Barbara C. Farhar, and Olin H. Foehner, Jr.; (3) Ray Jay Davis; (4) Lewis O. Grant, Harold D. Orville, Marcia Politovich, Roger F. Reinking, David Rogers, and Joseph Warburton; (5) Don A. Griffith, Marcia Politovich, James H. Renick, David W. Reynolds, and David Rogers; and (6) Conrad G. Keyes, Jr., Joseph A. Warburton, and James H. Renick. Most of these individuals were involved with the Climate and Weather Change Committee of the Irrigation and Drainage Division of Management Group D of ASCE.

This current edition or Revision of the Manual was produced by the author(s) listed within each section and approved for publication by a majority of the Atmospheric Water Management (AWM) Standards Committee (SC). The members of the EWRI Revision of Manual #81 Subcommittee were: Conrad G. Keyes, Jr. (chief editor) and co-editors: Bruce A. Boe, George W. Bomar, Robert R. Czys, Thomas P. DeFelice, and Don A. Griffith. The other reviewers from the AWM SC and/or the Weather Modification Association included Arie Ben-Zvi, Joseph H. Golden, Thomas J. Henderson, Maurice D. Roos, and Joseph A. Warburton. The final reviewers from the "blue-ribbon" group for the EWRI Standards Development Council included Darin W. Langerud, Paul L. Smith, Mark E. Solak, and William L. Woodley.

DEDICATION

This Manual is dedicated to five of the original co-authors of the 1983 and/or 1995 versions of the Guidelines. These individuals each made significant contributions to the cloud seeding to augment precipitation community during the many years of their professional lives and served ASCE as dedicated volunteers during many years of the development and publication of this subject.

RAY JAY DAVIS passed away August 10, 2000, at his home in Provo, Utah. Ray received a B.A. from Idaho State University in 1948, a J.D. from Harvard Law School in 1953, and a L.L.M. from Columbia Law School in 1956.

In a personal piece concerning his life, he wrote, "By profession I am a teacher. Few joys can equal the thrill of sharing learning with others and of watching their growth and development come about. I teach law. There is no subject other than the gospel more exciting to teach. I firmly believe that laws are 'those wise restraints which make men free.' I am proud of the role that I have had in making our legal system function."

An academician throughout his 45-year legal career, Ray Jay was a Professor of Law at Brigham Young University from 1979 until his retirement in April 2000. He also taught law at the University of Arizona (17 years), Temple University, and the University of Arkansas.

His research career was primarily devoted to studying and writing about the legal rules that govern, or should govern, the appropriation and use of water, particularly water contained in the earth's atmosphere. He served as chair of a monumental project undertaken by the ASCE to produce a model state water code to be transmitted to all 50 state legislatures with a recommendation for adoption, and to be published abroad as a law reform source in foreign countries. He was also the author of the Legal section of the first edition of ASCE Manual No. 81 as well as the initial version of the Guidelines in 1983.

Ray served as the chair, a member, a principal investigator, or an adviser to countless committees, to governmental agencies of different states, and to agencies of the federal government. He represented the United States at the United Nations Conference on International Legal Principles for Weather Modification. He made presentations at conferences in foreign countries and served as an adviser on the legal ramifications of cloud seeding to nine western and mid-western states. Some of his writings have been translated into French, Russian, and Spanish. A prominent legal treatise states, "Professor Ray Davis is the leading figure on weather modification law" (Robert Beck, *Water and Water Rights*, Vol. 2, Section 3.04[a]). His resume lists a total of 193 published items, including nine books and 20 chapters in books and treatises.

He was especially proud of authoring the *Arizona Workers Compensation Handbook,* the draft *Model State Water Allocation Code,* and the textbook *Law In Action: American Government.* Ray Jay was an active member of the Weather Modification Association and he served as its 'legal eagle.' He received the WMA's Thunderbird Award in 1978 at the association's annual meeting held in Tucson, Arizona.

ROBERT D. ELLIOTT, a true pioneer in purposeful weather modification, died of a stroke at his home in Santa Barbara, California on April 5, 2002: Bob was 87 years old. He graduated from the California Institute of Technology in 1937 with an M.S. in Meteorology. During World War II, Bob was a Naval Aerological Officer based in Washington, D.C. This group was responsible for the preparation of all weather forecasts for U.S. Naval operations, including the preparation of forecasts for the D-Day invasion. During this period he developed a storm typing system that is still in use by some meteorologists today.

In 1950, Bob helped form North American Weather Consultants (NAWC) in Pasadena, California. In 1951 NAWC was divided, with Bob leading the weather modification division, which was relocated to Santa Barbara, California. NAWC activities included weather forecasting, both short range and long range, in addition to weather modification.

One of the first weather modification projects undertaken by NAWC under Bob's direction was the design and implementation of a winter orographic cloud seeding program for the upper San Joaquin River Drainage located on the west side of the southern Sierra Nevada of California, which was supported by the Southern California Edison Company. Of historical note is the fact that this program has continued in an uninterrupted fashion to the present and is the longest continually operated weather modification program in the world.

The first *Journal of Weather Modification*, published in 1969, contains the following discussion on the background of the Weather Modification Association: "On April 4, 1951, Messrs. Stuart Cundiff, William Lang, Eugene Bollay, Robert Elliott, John Battle, and E.C. Hartman, met during a luncheon at the Mission Inn in Riverside, California. The object of this meeting was to discuss possible methods of organizing and controlling cloud seeding operations and evaluations in California for purposes of raising the standards with respect to those engaged in the business of weather modification." Bob was appointed Treasurer of the organization with the suggested name of Artificial Precipitation Operators Association. At a subsequent meeting on April 17, 1951, the name of the organization was changed to Weather Control Research Association (now the Weather Modification Association). Bob served as President in 1951 and 1952, and Vice President from 1957–1959. Bob was honored by the WMA in 1973 as the recipient of the first Thunderbird Award and in 1978 he was selected as the third recipient of the Schaefer Award.

Bob participated in a number of landmark weather modification research programs throughout his professional career. Among these were the early Santa Barbara experiments conducted in Santa Barbara County, the Bureau of Reclamation's Colorado River Basin Pilot Project, and the Sierra Cooperative Pilot Project. One of Bob's interests through his involvement with these research programs was the development of computerized targeting models that could be used to calculate the transport of cloud seeding materials, their interaction with the cloud microphysics, and the resultant fallout of seeded precipitation. Bob was heavily involved in the development of a model that could be used in real time to help meteorologists predict this sequence of events.

The American Meteorological Society (AMS) in 1961 honored Bob with the presentation of the Award for Outstanding Contributions to the Advance of Applied Meteorology. He was elected a Fellow of the AMS and was a member of the original Board for Certified Consulting Meteorologists. He served on a number of committees, including ones organized by the American Society of Civil Engineers that, among other activities, developed the guidelines on cloud seeding in 1983. Bob was also a member of the American Association for Advancement of Science, the American Geophysical Union, and Sigma XI.

OLIN H. FOEHNER, JR., who served as the first Director of the Sierra Cooperative Pilot Project (SCPP) of the Bureau of Reclamation of the U.S. Department of the Interior (USDI), and who was the original ASCE author of Section 2 of these guidelines, was lost at sea while scuba diving near St. Martin in the West Indies, May 27, 1983. During his active and productive career, Olin was a strong proponent of weather modification research and operational programs at the international level. Olin exercised a leading role in the planning and design of SCPP from its early stages until the spring of 1981 when he was reassigned as Director of the Colorado River Enhanced Snowpack Test (CREST). During his time as SCPP Director, the project moved from the initial planning to the design phase, the project's Auburn field office was established, the Skywater X Conference on the SCPP Design was held, the Sierra Ecology Project was initiated in cooperation with the Forest Service Pacific Southwest Forest and Range Experiment Station, and numerous other cooperative activities were initiated with the states of California and Nevada, various universities, and the private sector. A project public involvement program with active participation of members of the Citizens Council was also created.

Olin's energy, his dedication to the long-term Bureau's Skywater objectives, and his appreciation for new ideas contributed immeasurably to the progress and success of SCPP. The Division of Atmospheric Resources Research, the Bureau, and many colleagues miss both his expertise and his good humor.

DONALD ROTTNER co-founded the OPHIR Corporation, a research and instrumentation company that focused on the atmospheric sciences. Don was President of OPHIR at the time of his death in Lakewood, Colorado on May 23, 1995, about four months after the final editing of the ASCE Manual No. 81, of which he was one of the co-editors. In January 1980, Rottner started the OPHIR Corporation and was granted patents by the U.S. Patent and Trademark Office bearing his name as a co-inventor.

In June 1963, the Air Force transferred Rottner to the University of Wyoming as a student and officer trainee. He received his B.S. in civil engineering in 1965, was commissioned in December 1965 and began service as a bioenvironmental engineer in the Biomedical Sciences Corps of the Air Force. He left the military to enroll in the Department of Atmospheric Sciences at the University of Wyoming in September 1969. After receiving an M.S. in 1971, Rottner joined the staff of New Mexico State University, where he worked as an assistant project engineer on a cloud seeding project conducted by the NMSU Department of Civil Engineering. He joined the Division of Atmospheric Resources Research of the Bureau of Reclamation in June 1972 and he became a professional member of the AMS that same year.

Working on Project Skywater, Rottner made significant scientific contributions. He was responsible for assembly, quality control, and archival of the data collected by a five-year, multimillion-dollar weather modification program in southwestern Colorado. The data management program maintained a huge database that included digital radar, satellite imagery, cloud physics measurements, rawinsondes, pibals, acoustic sounders, ground-based radiometers, precipitation networks, and ice crystal habits and concentrations. He used his expertise with OPHIR Corporation and his past experience in weather modification to become one of the co-editors of the ASCE Manual No. 81 in 1995 and he influenced at least two other employees of the OPHIR Corporation to be heavily involved with the same publication.

JOSEPH A. WARBURTON suddenly and peacefully passed away at home on April 30, 2005, in Reno, Nevada. His contributions to the scientific community, Masonic fraternity, and humanity were larger than life.

Joe served in the Australian Imperial Forces during World War II. Following graduation from Goulburn High School in Goulburn, NSW, Australia, in 1946 he attended the University of Sydney, graduating with honors in physics and mathematics. Advanced studies in radio astronomy and the physics of the lower atmosphere led to a Master's Degree and Ph.D. at the University of Queensland. He was employed at C.S.I.R.O. in Sydney. Dr. Warburton established 'The Warburton Family Science Award' at Goulburn High School to provide scholarships to outstanding science students.

In mid-1965, Warburton was appointed to a senior scientist position at the Desert Research Institute. From 1969 to 1970 he served the University System as the President of DRI and later as the Executive Director of the Atmospheric Sciences Division from which he retired in 1993, the University Board of Regents awarding him Emeritus status. At the time of his death, Joe was working on a weather modification program he developed for the Snowy Mountains Hydro-electric Authority in Australia, in addition to writing a book titled *The Science of Weather Modification*.

Dr. Warburton's scientific work is described in over 120 papers published in scientific journals in the United States and other countries. He conducted research projects in Antarctica, France, Greenland, Switzerland, Canada, China, Australia, Morocco, Saudi Arabia, Iran, and Spain. Joe was appointed a Fellow of the Australian Institute of Physics, a member of the American Meteorological Society, Secretary/Treasurer of the North American Interstate Council on Weather Modification, a member of the Antarctican Society and an alumnus of the University of Queensland. His scientific awards include the Antarctic Service Medal, the Vincent L. Schaefer Scientific award for outstanding original contributions in the field of weather modification, and his appointment as Visiting Fellow at the Australian National University in Canberra, Australia in 1996. He was recently honored for his work in the Antarctic by having a landmark named after him—"Warburton Ledge," located 4 miles east of Mount McClintock in the Britannia Range, Antarctica.

SECTION 1
EXECUTIVE SUMMARY

Thomas P. DeFelice,[1] Ph. D., WMA CO, M. ASCE; and Conrad G. Keyes, Jr.,[2] ScD, P.E., P.S., D.WRE, WMA CM/CO, Hon. M. ASCE (Life), F. NSPE (Life)

1.1 INTRODUCTION

Modern cloud seeding technologies may be successfully applied to help resolve community issues, and have been for over 50 years. Recent technological and scientific advances, along with contemporary socio-economic problems, strengthen the impetus for seeking applications of modern cloud seeding technologies that could benefit our society, primarily in regions where additional precipitation is viewed as an economic asset. The augmentation possible is fractional, and where successful, may be in the range of 5% to 20% (Elliott et al., 1995). However, this much additional rain water over the farm belt could benefit agriculture, and over mountainous terrain could benefit the hydroelectric power industry, municipal water supply, and irrigation interests. The technology is not without limitations, which must be recognized and incorporated into decisions regarding its use.

It remains necessary to develop public consensus within an intended target area, because the smallest possible scale of treatment covers several hundred hectares or several million square meters. Many farmers might benefit from enhanced precipitation, while others, such as those in a farm area having mixed crops, might not. In a mountainous region where hydroelectric power generation would be greatly benefited, traffic over mountain passes might be impaired, while ski resorts might be aided (Elliott et al., 1995). Rational assessments of each implication require scientific studies, which to date all suggest that developing the operational application of cloud seeding technologies can help mitigate the aforementioned impacts, provide sustainable water supplies, help reduce airborne hazards, and even improve evaluation methods for operational

[1]Program Manager, Science & Technology Programs, I.M. Systems Group, Rockville, MD 20852.

[2]Chair, ASCE/EWRI Standards Development Council and Past Chair, ASCE/EWRI Atmospheric Water Management Standards Committee and Co-Editor of the 1995 ASCE Manual No. 81, PO Box 1499 Mesilla Park, NM 88047.

activities. Silverman (2001a, 2001b) suggests that it might not be prudent to wait until drought to employ cloud seeding technology, and that cloud seeding technology should be implemented as an integral part of an overall management strategy for a watershed or region.

1.2 WHY SEED CLOUDS?

Clouds form as part of the hydrologic cycle, which describes the processes of evaporation of water into (and within) the atmosphere, condensation, precipitation that runs along and within the lithosphere into water bodies only to repeat the cycle again. The efficiency of the precipitation process may be enhanced. Precipitation efficiency is defined as the fraction of condensed water vapor that ultimately reaches the ground. The precipitation efficiency of a thunderstorm, for example, has been approximated as 19% (e.g., Houghton, 1968).

Cloud seeding technology can help facilitate some components of the water and energy cycles, which are key to dealing with many present and potential future scientific, environmental, and socioeconomic issues. A common misconception is that cloud seeding robs "Peter's" rain to water "Paul's" land. Properly conducted glaciogenic seeding increases the precipitation efficiency, resulting in more total rainfall that falls over a broader area compared with the unseeded case. Hence, cloud seeding benefits both "Peter" and "Paul." Obviously this is a good thing if you are a farmer. In the case of precipitation enhancement, the weather modification technology (cloud seeding) applied to a thunderstorm boosts the precipitation efficiency. There have been significant advances in our understanding of the hydrologic and energy cycles. These advances, when combined with improved scientific understanding, provide a better picture of when and where the atmosphere or cloud can most likely benefit.

There are determinable benefits from cloud seeding to enhance precipitation efficiency within a cloud. These benefits manifest themselves in terms of increased hydroelectric power and agricultural production, salinity reduction, and strengthened ski industries, while water supplies are improved for fish and wildlife, recreation, municipalities, and industry. Thus, while the primary motivation for cloud seeding may be economic, there are clearly other potential benefits to the public and private enterprise. Elliott et al. (1995) reported rainfall increases of 10% to 40% (compared to normal rainfall) would increase corn and soybean yields by 4% to 20% if natural rainfall was near normal. The direct beneficiaries in winter mountain snowpack projects include those using the resultant excess streamflow for hydroelectric power generation, irrigation water, and ski resorts. It has been estimated that a 1% to 2% increase in rainfall (compared to normal rainfall) would pay for such cloud seeding projects.

1.3 APPROACHES AND RESTRICTIONS TO SEEDING CLOUDS

Awareness of public concerns, a responsive and well-guided public involvement program, a corresponding decision process, and ongoing evaluation of both the direct and indirect effects will provide many appropriate checks and balances as cloud seeding programs are conducted. However, the risk perceived by stakeholder groups and the public will be very important in whether the community accepts a cloud seeding project.

There are potential risks associated with the use of cloud seeding technology. Social, environmental, and economic factors will determine whether or not a cloud seeding program is accepted. Risk-benefit assessments in each of these categories, relative to alternatives for providing more water, are appropriate. Legal guidance and restraints concerning cloud seeding assure fair balance between opportunities to advance individual and group desires and concerns and the need to consider the rights of the remainder of society.

The experiences from more than 50 years of seeding clouds, relevant lessons learned exercises, conferences, discussions, and general public forums have generated, among other things, environmental issues, including potential effects on cultural resources, erosion rates, duration of snowmelt, and contributions to the "greenhouse effect." The developments of Environmental Impact Statements (EISs) and Reports have been necessary in some cases. Some early environmental concerns focused on the seeding agent, i.e., the ice crystal nucleant, AgI (silver iodide). Heavy metals occur in nature and residual silver from seeding is normally produced in concentrations far below toxic levels. This is only one of many environmental aspects of cloud seeding. It should be recognized that effects of added water on the environment could be negative as well as positive. Organizations that undertake operational cloud seeding should be prepared to invest the considerable time and costs of preparing EISs, especially if federal funds are used, and to consider the subsequent costs of environmental monitoring during operations (e.g., Elliott et al., 1995).

The section on legal considerations applied to atmospheric water management or cloud seeding has been adapted from Davis (1995) with adjustments to reflect recent developments pertaining to legal implications of the use of cloud seeding technology. Legal considerations apply to atmospheric water development implementation decisions in the same way that they apply to development of any other part of the hydrologic cycle. The ability of the federal government to impact weather modification policy resides within various appropriation acts that make government funding available for research in, and development of, cloud seeding technologies. The only federal statute governing weather modification has to do with reporting activities; practically all of American law specifically targeting weather control activities rests with the states.

As a means for avoiding misapplication of the technology by poorly qualified individuals, or by groups focusing narrowly on special interests' benefits, or in uncontrolled, unmonitored, or conflicting projects, a legal system has gradually been developed for controlling the application of the technology. Many states have regulatory laws in place. Licensing, permitting, and reporting may be required. A U.S. law requires that persons carrying out weather modification activities report them (Elliott et al., 1995). An appendix in Section 3 of Davis (1995) provides one list of state statutory and regulatory references on weather modification.

1.4 SCIENTIFIC BASIS FOR CLOUD SEEDING

Cloud seeding technology is based on sound scientific principles. Section 4 describes the scientific basis necessary to provide an adequate understanding of how and why precipitation enhancement might be achieved. There is still much to learn about how clouds naturally produce precipitation, despite the significant advances made in recent years with regard to the amount and quality of empirical data, theoretical development, and instrumentation dedicated to this natural phenomenon. The fundamental scientific premise has been that a cloud's precipitation efficiency can be increased, or that a cloud's vertical development could be enhanced. The result is a cloud with a more efficient precipitation process (i.e., a more productive cloud). Initial experiments in the middle 1940s under the direction of Nobel Laureate Irving Langmuir at the General Electric Laboratories in Schenectady, New York, and the 1946 seeding field test by Vincent Schaefer in which dry ice was dropped into a stratiform cloud deck, were the first examples to support this premise. The dry ice acted very quickly after entering the supercooled water droplet cloud to transform the cloud hydrometeors into millions of tiny ice crystals that grew and fell from its base, thereby increasing the efficiency of the precipitation process, and leaving behind a distinct clearing in the cloud deck. This stratocumulus cloud evidently did not have many available naturally occurring ice forming nuclei to initiate the precipitation process.

Later, Bernard Vonnegut of the G.E. Laboratories discovered how to produce ice embryos by introducing a swarm of minute silver iodide (AgI) smoke particles into a supercooled cloud. Their structure is similar to that of ice, and water vapor deposits on them to form ice. Subsequently, other investigators discovered other nucleating agents. Presently, AgI produced in complexes with other chemicals remains the chief agent in use, although dry ice and some organics are occasionally used. Elliott et al. (1995) provided details of cloud seeding agents and ice crystal nucleation not covered in Section 4.

Scientific studies of ice crystal nucleation also yielded questions about the details of how nature produces precipitation. It is now recognized that clouds suitable for seeding are supercooled, relatively free of ice at critical times in their evolutions, and have appreciable natural dynamical forcing (Elliott et al., 1995). Clouds not meeting these criteria have little chance of producing precipitation and cannot be usefully seeded for this purpose. These studies have also led to the need for technologically advanced tools, designed to measure, numerically model, and verify cloud processes, the dispersion of seeding material, and the effect of the seeding material on the cloud precipitation processes. These technologies would also be useful for evaluating cloud seeding potentials and effects. All tools have their limitations and these should be included in any risk-benefit analysis, but those for cloud seeding have improved dramatically in recent years.

1.5 THE CONDUCT OF CLOUD SEEDING OPERATIONS

Seeding operations are conducted for a number of reasons, besides the fact that cloud systems are inefficient at producing precipitation that reaches the ground. The principal elements of cloud seeding operations are the cloud with a potential to have its precipitation efficiency augmented, seeding material selection and its delivery and dispersion within the cloud volume, the resulting cloud physical, dynamical, and microphysical transitions to stimulate additional precipitation, meteorology associated with the cloud system, and the fallout of precipitation. Section 5 focuses on the seeding material selection, its delivery and dispersion within the cloud, and the resulting dynamical cloud effects. That is, the how-to-do-it section.

The most commonly used method for producing artificial ice-forming nuclei is a seeding device or "generator" that vaporizes AgI in solution with acetone and other chemicals, emitting numerous tiny (0.01 to 1.0 μm) silver iodide (AgI)-containing nuclei. Different AgI-complexes nucleate clouds at different rates and by different microphysical mechanisms, so appropriate selection is important. Cloud temperature is a governing parameter. The AgI-containing nuclei generally become active ice nucleants at air temperatures near and colder than -4°C. Laboratory tests of the Weather Modification Inc. (WMI) wingtip generators with North Dakota AgI-acetone formulation yielded approximately 10^{14} ice particles per gram of AgI produced at -10°C (DeMott, 1997). A second popular method is to drop dry ice pellets into a cloud that contains supercooled water droplets. The dry ice method generally yields 10^{12} or more ice crystals per gram of dry ice, and is effective at air temperatures as warm as about $-1°$ to -2°C.

Effective delivery and dispersal of the seeding agent from the source through the recipient cloud requires meticulous planning for optimal implementation during operations. The type of generating system employed and its mode of operation depend upon the type of cloud systems requiring treatment. It also depends on making full use of available historic meteorological data and ancillary data where appropriate, time of operations (winter, summer), target area size, topography, accessibility, funding availability, and other project-specific aspects. Ground-based systems are most useful for wintertime projects in mountainous areas. They have limited utility in summertime projects. Aerial dispensing systems are ideally suited to summertime cumulus seeding either at cloud base, in cloud, or at cloud top. Both silver iodide and dry ice can be dispensed aerially; silver iodide and liquid propane can be dispensed from the ground, but dry ice cannot.

All projects require monitoring of seeding agent delivery and dispersal, as well as evaluations to quantify their success. The evaluations often rely on statistical techniques because the cost of direct monitoring technologies can be high. Evaluations must include physically or chemically based approaches to quantify success. Seasonal streamflow or snow course, precipitation gauge, wind sensor, upwind rawinsonde, and modernized weather radar data (e.g., Elliott et al., 1995; DeFelice, 1998; ASCE, 2004) might all be useful to evaluation. Instrumentation provides needed input data for real-time decisions, such as the forecasting of probable seeding opportunities, the determination of seedable situations, conducting seeding operations, and the exercise of project suspension criteria. Instrumentation can also provide data for post-project assessment of the probable effects of cloud seeding based upon critical parameters, such as precipitation or streamflow.

Project planners should bear in mind that the hygroscopic flare method is relatively new and is not yet used as widely as the AgI complexes, although it has shown considerable promise (Cooper et al., 1997; Mather et al., 1996, 1997). Additional experimentation utilizing this technique has been conducted in Mexico for rain enhancement (Bruintjes et al., 1999). Future experimentation needs to demonstrate that this technique can increase precipitation over a fixed target area for a significant period of time (e.g., a summer convective season).

The advances in technologies used to enhance precipitation or to monitor its success have allowed research project investigators to move away from purely statistical evaluations toward physical evaluation derived from direct observations of the seeding material delivery to clouds, the resulting physical response, the precipitation fallout, and the chemistry of seeded and unseeded precipitation that has reached the ground. The vastly improved numerical models that simulate cloud processes complement many new remote and in situ atmospheric and cloud-sensing technologies (e.g., sensor for continuously profiling the wind and temperature). Measuring devices once considered for use in purely research projects are gradually moving into operations and should be reviewed by the would-be cloud seeder and used where affordable for both real-time guidance and post-project evaluation.

1.6 HOW TO INITIATE A CLOUD SEEDING PROJECT

This section has been adapted from Keyes et al. (1995) with major adjustments to reflect recent developments pertaining to the initiation or implementation of a cloud seeding project. The implementation of a cloud seeding program is an area where environmentalists, meteorologists, and water planners must work together to safely and effectively yield additional water resources to a target area. This final section provides readers with a blueprint for arguably the most difficult component of a cloud seeding project, i.e., its implementation. It suggests what some of the basic questions should be and then proceeds to methods of getting the answers.

Each scientist, engineer, and planner associated with a cloud seeding project must consider the overall need for the precipitation augmentation, know the feasibility, and control of the program before implementation is initiated. Other important considerations: approaches to augmenting precipitation, and, once underway, program performance and continued management, as well as evaluation and lessons learned.

Criteria that administrative agencies consider for permit granting include project personnel and their field experience, seeding agents and modes, equipment, target area, operational plan, safeguard criteria, information gathering and evaluation plan for projected impact of seeding, and contract and cost information. Advisory boards or committees of experts often aid in the permit decision.

This section covers such topics as Initial Program Assessment and the factors governing implementation; Needs and Goals including the origin of need and program justification, political and/or institutional justification; the feasibility study, including program expectations and its objectives; program definition, which defines seeding modes and agents, an evaluation plan and quantification of findings; program control as in seeding decisions, data collection and access, seeding suspension criteria; and program management including a lessons learned exercise.

1.7 CONCLUSIONS

There are determinable benefits from cloud seeding to enhance precipitation efficiency within a cloud. These benefits may manifest themselves in terms of increased hydroelectric power and agricultural production, salinity reduction, and strengthened ski industries, while water supplies are improved for fish and wildlife, recreation, municipalities, and industry. Recent evaluations have shown that precipitation increases from 5% to 20% can be achieved through effectively operated cloud seeding programs. The startup and continuation of a cloud seeding program will most likely depend on the viability of the technology, the perceptions of the benefits and liabilities as derived from the whole environmental, social, economic, and legal process, the diligence with which effects are monitored, and how well public involvement is maintained.

1.8 REFERENCES

American Society of Civil Engineers (2004). *Standard Practice for the Design and Operation of Precipitation Augmentation Projects.* ASCE/EWRI Standard 42-04, ASCE, Reston, VA, 63 p.

Bruintjes, R.T., Breed, D.W., Foote, G.B., Dixon, M.J., Brown, B.G., Salazar, V., and Rodriguez, H.R. (1999). "Program for the augmentation of rainfall in Coahilla (PARC): Overview and design." *Preprints, Seventh WMO Scientific Conf. Wea. Modif.*, Technical Document No. 936, World Meteoro. Org., Geneva, 53–56.

Cooper, W.A., Bruintjes, R.T., and Mather G.K. (1997). "Some calculations pertaining to hygroscopic seeding with flares." *J. Appl. Meteorology,* **36**, 1449–1469.

Davis, R. (1995). "Legal Aspects of Weather Modification Operations." In *Guidelines for Cloud Seeding to Augment Precipitation,* Section 3. ASCE Manuals and Reports on Engineering Practice No. 81. ASCE, Reston, VA, 49–68.

DeFelice, T.P. (1998). *Introduction to Meteorological Instrumentation and Measurement,* Prentice Hall, Saddle Brook, NJ, 229 p.

DeMott, P.J. (1997). Report to the North Dakota Atmospheric Resource Board and Weather Modification Incorporated on Tests of the Ice Nucleating Ability of Aerosols Produced by the Lohse Airborne Generator—1997. Department of Atmos. Sci., Colorado State University, Ft. Collins, CO, 14 p and appendices.

Elliott, R.D., Keyes, Jr., C.G., and Reinking, R.F. (1995). "Summary." In: *Guidelines for Cloud Seeding to Augment Precipitation,* Section 1. ASCE Manuals and Reports on Engineering Practice No. 81. ASCE, Reston, VA, 1–7.

Houghton, H.G. (1968). "On precipitation mechanisms and their artificial modification." *J. Appl. Meteorology,* **7**, 851–859.

Keyes, C.G., Jr., Warburton, J.A., and Renick, J. (1995). "How to Implement a Cloud Seeding Program." In: *Guidelines for Cloud Seeding to Augment Precipitation,* Section 6. ASCE Manuals and Reports on Engineering Practice No. 81, ASCE, Reston, VA, 125–136.

Mather, G.K., Dixon, M.J., and deJager, J.M. (1996). "Assessing the potential for rain augmentation—the Nelspruit randomized convective cloud seeding experiment." *J. Appl. Meteorology,* **35**, 1465–1482.

Mather, G. K., Terblanche, D.E., Steffens, F.E., and Fletcher, L. (1997). "Results of the South African cloud-seeding experiments using hygroscopic flares." *J. Appl. Meteorology,* **36**, 1433–1447.

Silverman, B.A. (2001a). "A critical assessment of glaciogenic seeding of convective clouds for rainfall enhancement." *Bull. Amer. Meteor. Soc.,* **82**, 903–923.

Silverman, B.A. (2001b). Comments on "A critical assessment of glaciogenic seeding of convective clouds for rainfall enhancement." Reply. *Bull. Amer. Meteor. Soc.,* **82**, 2848–2849.

SECTION 2

SOCIETAL, ENVIRONMENTAL, AND ECONOMIC ASPECTS OF PRECIPITATION ENHANCEMENT BY CLOUD SEEDING

Conrad G. Keyes, Jr.,[3] *ScD, P.E. P.S., D.WRE, WMA CM/CO, Hon. M. ASCE (Life), F. NSPE (Life)*

2.1 INTRODUCTION

Sociologists, ecologists, economists, geographers, and political scientists, as well as atmospheric scientists, have addressed the complex issues that arise with the development and application of cloud seeding technology. Selected information exemplifying these aspects of cloud seeding is summarized in this chapter. The material is intended to stimulate appropriate consideration for the interdisciplinary factors that affect a successful cloud seeding program and/or projects.

2.2 SOCIETAL ASPECTS

The decision to adopt cloud seeding has historically focused on the question, "Can we do it?" However, as technical capabilities improve, a value question arises: "Should we do it?" (Sewell, 1969). Attempting to modify precipitation by seeding clouds is a product of people's continual search for means to manage their environment. Even in situations where cloud seeding offers potential economic benefits or environmental advantages, the weather can be purposefully managed only if those affected agree that it should be done (Borland, 1977; Farhar, 1977). Much of what actually can be done is governed by societal choice. The human dimensions of cloud seeding programs must be considered if the technology is to be effectively used (Sewell, 1966).

[3]Chair, Subcommittee for Revision of Manual ASCE #81, and Chair, EWRI Standards Development Council, P.O. Box 1499, Mesilla Park, NM 88047, (505) 523-7233, cgkeyesjr@zianet.com

Interest in cloud seeding has historically risen during dry periods and waned when rain and snow are plentiful. Most communities where cloud seeding has been carried out have accepted scientific experimentation, and some communities have actively sought operational projects. In a few communities, grassroots groups organized political opposition against cloud seeding in states such as Delaware, Colorado, California, Texas, and South Dakota. For proponents and opponents alike, once issues and attitudes toward cloud seeding are established, they tend to persist for a very long time (Farhar and Fitzpatrick, 1990). Therefore, the would-be cloud seeder or water manager may be served best by understanding insights from social science about public response to cloud seeding and introduction of new technology in general (Reinking et al., 1995).

2.2.1 Studies

Research on the social aspects has focused on public response to field projects where they have been proposed and introduced, and on decision processes regarding the adoption of cloud seeding. Four kinds of studies have been conducted: (1) surveys of citizen attitudes, opinions, beliefs, knowledge, and favorability toward cloud seeding (e.g., Haas et al., 1972; Larson, 1973; Farhar and Mewes, 1975; Krane, 1976; Farhar and Rinkle, 1977); (2) monitoring of project areas to determine the factors associated with acceptance and rejection of cloud seeding (e.g., Haas, 1974; Farhar, 1975a, 1976, 1977, 1978; Farhar and Mewes, 1975; Farhar and Fitzpatrick, 1990); (3) a technology assessment of hail suppression conducted by an interdisciplinary team (Changnon et al., 1977); and (4) a nationwide survey of weather modification experts (Farhar and Clark, 1978).[4] Also, the 1988–1990 study by Farhar and Fitzpatrick (1990) revisited four cloud seeding projects studied a decade or more earlier, and Walkinshaw (1985) reported on the economic and social impacts of long-term weather modification projects in the United States.

2.2.2 The Diffusion of Innovations and Cloud Seeding

The diffusion of any technology into a population follows a well-understood pattern (Rogers and Shoemaker, 1971; Rogers, 1983). The rate of adoption is more rapid for technologies adopted by individuals (such as hybrid seed corn or the birth-control pill) than for technologies

[4]The most extensive research on public response to weather modification was conducted at the Institute of Behavioral Science at the University of Colorado, Boulder, and later moved to a small private research firm in Boulder, Colorado. The work was conducted at the Hazards Assessment Laboratory, Colorado State University, Fort Collins.

adopted by communities (such as kindergarten or cloud seeding). In the former case, the innovation can reach saturation in as quickly as 5 years, whereas it took 50 years for kindergarten to be adopted by essentially all U.S. communities (Reinking et al., 1995).

The population usually can be represented as a normal curve based on the length of time individuals take to make the adoption decision (Reinking et al., 1995). At the leading edge of the normalized population curve, representing about 2.5% of the population, are the innovators. They adopt quite early in the process and reap the benefits of doing so ahead of others, i.e., the innovative farmers who were the first to adopt hybrid seed corn reaped the financial benefit of having higher corn yields well before their neighbors.

Early adopters, the next 13.5% of the population, accept the innovation soon enough to benefit from early acceptance; opinion leaders are usually found within this group. The innovators and early adopters are highly receptive to change relative to the specific idea or technology that they favor, but they do not necessarily support all new ideas or technologies. The same innovators would not necessarily favor cloud seeding, hybrid seed corn, and birth-control pills. Following these groups are the early majority (those still adopting the innovation on the early side of the mean time required by the population) and the late majority (those adopting somewhat more slowly than the mean). This intermediate, great majority, representing two-thirds of the population, neither actively promotes nor opposes the introduction of new technology (Rogers and Shoemaker, 1971; Dennis, 1980). The "laggards," comprising some 16% of the population, would prefer no change and are the slowest group to adopt. If a new technology does not work but is applied anyway, it is the laggards who are most likely to avoid any associated disbenefits. The opinion leaders (part of early adopters) shape the attitudes and actions of the early majority toward any innovation, including cloud seeding. This is why the opinion of community leaders is so important in the social acceptance of cloud seeding projects (Reinking et al., 1995).

2.2.3 Assessing Public Attitudes

The *perceived* value of cloud seeding, derived by weighing both potential opportunities and adverse effects, incorporates considerations beyond simple economics and the efficacy of the technology. The issues surrounding cloud seeding may reach deeply into social relationships and even into aesthetic and spiritual values, including concern about the risk of human intervention in weather processes. Fear of disasters that might be perceived as potential effects of cloud seeding, such as flooding, avalanches, or ecological calamities that might carry property and social losses, also shapes attitudes (Sewell, 1966; Larson, 1973; Farhar, 1977;

Dennis, 1980). There are, and most likely always will be, those who are convinced that it is possible to modify the weather, and those who believe it is a fraud (Sewell, 1966), irrespective of the scientific basis (Reinking et al., 1995).

Scientifically conducted surveys can measure the distribution in factors that affect the social acceptability of projects. Examining the range of social factors that can come into play provides some perspective on this. Attitudes toward weather modification in South Dakota, Colorado, and Florida were measured using survey items that measured (1) favorability to the technology, (2) beliefs that cloud seeding was effective in increasing precipitation and suppressing hail, (3) concern about risk, (4) credibility of sources of information about the cloud seeding program, (5) knowledge about the program, (6) evaluation of the cloud seeding program, and (7) preferred decision making and funding sources (Farhar, 1975a, 1975b).

Belief in the effectiveness of the technology, favorable attitudes toward it and toward science, and low concern about risk were statistically most highly correlated with positive public assessment of operational seeding programs. Belief in efficacy emerged in this and other research as the most powerful predictor of societal program evaluation outcomes. Information on the efficacy and secondary effects of cloud seeding is commonly and appropriately sought from the technical experts. Scientists themselves have divergent views on the efficacy (Farhar and Clark, 1978). Although this is the nature of scientific probing, public disillusionment can be fueled by scientists who disagree in public about the facts and the uncertainties (Lambright, 1972; Changnon and Lambright, 1990). Conversely, problems may occur if operators do not openly state the uncertainties. More recent AMS, ASCE, WMA, and WMO statements on atmospheric water management have been presented in the recent standard practice documents on hail suppression and precipitation enhancement (ASCE, 2003, 2004, and 2005). Some of the details of these statements are provided in Section 5.4 and other reports by the National Research Council of the National Academy of Science on weather modification are under review at this time.

The South Dakota experience with cloud seeding is significant because the first state-sponsored weather modification program began there in 1972. At its height, the program included about 60% of the land area of the state at a cost of approximately $1 million each growing season. Surveys of the South Dakota public from 1972 to 1975 found the majority were favorable toward the program (Farhar, 1973; Farhar, 1975b). South Dakotans favored cloud seeding because they believed it would help farmers in a state where agriculture is the mainstay of the economy. Only after an opposition group organized, and the legislature failed to appropriate funds, did majorities fail to favor the program (Reinking et al., 1995). In contrast, North Dakota has maintained an operational program with public support, and some opposition, for over 40 years.

When a weather modification project has local sponsorship and strong local support over several years, even a large-scale negative weather event (such as a flash flood disaster) in the presence of cloud seeding may not produce organized protest. For example, findings from South Dakota show that the majority of respondents did not attribute the Rapid City flood to cloud seeding that had been carried out in the area. Even those who thought cloud seeding was to some extent responsible for the flood did not organize against it. A committee of scientists released an official report 18 days after the flood stating that cloud seeding was not its cause; this may have alleviated some citizen concern (Farhar, 1974).

2.2.4 Assessing Community Dynamics

The decision to adopt cloud seeding is most often made by organizations and communities, not by individuals. Therefore, public opinion is but one factor in a community where this decision is being made. Furthermore, although a number of communities adopted cloud seeding during the 1970s, most of these projects have been discontinued since that time. This can probably be attributed to early overselling and consequent unrealistic expectations of a technology still in its infancy. Cloud seeding adoption by communities has not followed the bell-shaped curve.

Public acceptability is not identical with unanimous agreement. Public acceptance means that sufficient majority support exists to move a project forward without serious social polarization. Systemic (community-level) variables relevant to acceptance of cloud seeding (Haas et al., 1972; Weisbecker, 1974; Mewes, 1977) include the following (Reinking et al., 1995):

1. Environment and conditions: Climate, water supplies, economic activities, topography, and population.
2. Structure and process: Institutions, organizations, power elites, stakeholder interests, and relationships among groups.
3. Interaction between the project community and communities downwind and downstream.
4. The stream of events: Contingency events and planned action.

Stakeholder groups—the range of associations; state, local, and tribal governments; and individuals with economic or domain interests—may be affected by the proposed action. Difficulties in the application of a cloud seeding technology can arise if certain groups can be identified as economic losers (Haas, 1974). The direct benefits of augmenting snowpack are in some cases realized downstream, rather than in the impacted target area; in one such case, some project area residents were opposed to cloud seeding because of "a resentment of outside politicians who make their winters longer and their weather-related problems worse, ... and the perception that the controlling authority will not be influenced by the wishes of the project area residents" (Weisbecker, 1974). In some agricultural areas, where direct impacts and benefits of cloud seeding are collocated, the greater the affiliation of the cloud seeding program with some farm organizations, the more favorable the attitude toward the program will be (Larson, 1973). Thus arises the issue of trust. Closely related is the issue of control. The desire is always strong to participate in, or have trusted representatives participate in, decisions that affect stakeholder interests.

The Sierra Cooperative Pilot Project (SCPP) offers another relevant case study. Agriculture, utilities, water districts, lumbering, cattlemen, and recreational interests use the Sierra Nevada's most valuable asset, water. SCPP's primary purpose was to develop and refine an effective snowpack augmentation technology that was socially as well as environmentally acceptable. In preparation for SCPP, the sponsoring U.S. Bureau of Reclamation and the State of California conducted 21 meetings in California and Nevada communities during 1974 to inform the public and to involve them in the project's planning process. Reclamation then sponsored a societal assessment of citizen and organizational response to the proposed project (Farhar and Rinkle, 1977). The assessment concluded that the proposed project appeared to be acceptable; however, diversity of weather modification needs and some fear of local disbenefits characterized the area. "The potential for opposition (was) definitely present, both at the systemic and individual levels" (Farhar and Rinkle, 1977). With a recommendation from the researchers, SCPP managers established a citizen advisory committee in January 1978. The managers attended numerous public meetings and distributed a monthly newsletter to interested organizations and individuals throughout the area.

After seeding began, some opposition was expressed over concerns about increased snow-removal costs, greater flood and avalanche hazards, reduced ski business resulting from excessive snow, increased road wear from tire chains, and general property damage. Project managers met with county supervisors and other interested organizations in the area, discussed their concerns with them, and successfully abated further opposition. SCPP also used project suspension criteria in cases of heavy snowfall and avalanche danger to mitigate these objections.[5]

---

[5]The cost-effectiveness of the snowpack augmentation programs when suspension criteria are in place still remains to be evaluated. That is, if an insufficient number of storms remain eligible for seeding operations, the program could become non-cost-effective.

These actions of involvement and responsiveness on the part of SCPP project managers, over a period of years, decreased stakeholder and public concerns. Despite the presence in the area of several factors that could have contributed to the organization of opposition, the project enjoyed 10 years of experimental cloud seeding data collection in an atmosphere of public acceptance (Reinking et al., 1995).

2.2.5 Decision Processes

Knowledge derived from the social science research in precipitation modification has helped and can help the involved parties communicate effectively and design strategies and policies that address public concerns fairly, allowing decisions to be made intelligently and responsively. So, sociologically, what are the most important factors affecting decisions to start, continue, or discontinue use of cloud seeding technologies (Reinking et al., 1995)? One social study [hereafter, the Colorado State University (CSU) study] analyzed community responses and changes in response to the application of cloud seeding technology in four areas over a 15-year period (Farhar and Fitzpatrick, 1990). Much of the following is based on the findings of the CSU study. These findings should not be regarded as universal but they do offer considerable insight. According to the study's analysis, predominant (albeit non-exclusive) factors that should be considered when deciding about cloud seeding are:

1. Built-in safeguards.
2. Local economic benefit.
3. Scientific evidence of effects.
4. Cost-effectiveness.
5. Drought conditions.
6. Compensation for disbenefits.

Clearly, a responsive approach toward stakeholders and public concerns can lead to a positive attitude toward project management (Farhar and Fitzpatrick, 1990; Keyes, 2002). Local control consistently has been found to be the preferred form of decision making on cloud seeding (Farhar, 1977). Satisfaction that the factors above are taken into account leads to community satisfaction with the decision, whether the project is accepted or not. Public involvement will increase project costs, and "educating the public" does not necessarily lead to acceptance (Farhar and Fitzpatrick, 1990). However, a public involvement program, tailored to requirements of an individual project, is the approach social scientists recommend to avoid community polarization and enhance the probability of community acceptance. This includes public information programs, citizen advisory committees, mechanisms for listening to constituency opinion, and a compensatory mechanism for disbenefits if these are supported by adequate information. Research that will answer questions raised by citizens and organizations may also be appropriate. Providing trained project personnel to work with community concerns may be appropriate (Reinking et al., 1995).

Formal decision processes that protect public interests and respond to public concerns lead to greater community satisfaction with the decision outcome. Figure 2-6 of Reinking et al. (1995) illustrates one concept of the stages in the weather modification decision process; factors affecting community response to cloud seeding are listed. All these community response factors feed into the decision-influencing factors listed above. According to the CSU study, the primary factor that determines whether cloud seeding technology will be accepted and applied is the informal influences in the project area at hand.

If the informal power brokers in the community favor its use or continuance, then a project will tend to go forward. . . . Those (power brokers) left unconsulted, uninformed, their concerns unattended to, become the seedbed of opponent action. The second factor . . . is the relative advantage (Figure 2-6) that adopting cloud seeding provides the community's members, as compared with not doing so or with doing something else. Cost-effectiveness and benefits to the local economy without undue risks (guaranteed by adequate safeguards) are essential to implementing and continuing projects. These benefits must be perceived by the local people, not merely asserted by proponents, scientists, or officials (Farhar and Fitzpatrick, 1990).

The safeguards might be suspension criteria for operations when snowpack is some defined level above normal or when avalanches are predicted, for example. Compensatory mechanisms for clear liabilities have appeared to be more important in winter than in summer projects.

Observation [Figure 2-6 of Reinking et al. (1995)] of the project's effects is the third key factor. Community members with interest will observe the effects of seeding, and they will come to conclusions about what cloud seeding is doing to their weather. If scientists bring sound evaluation data forward, this will carry some weight. However, the effects that the people themselves experience will be decisive. These observations are tied to belief in efficacy. If community members attribute (perceived or actual) positive effects—more rainfall and less hail for agriculture and more snow for ski areas—to cloud seeding, then project continuance is more likely to occur (Farhar and Fitzpatrick, 1990).

This indicates that it would be wise for those introducing the cloud seeding technology—the broker/change agent in Reinking et al. (1995)—to provide community representatives with observational experiences, by means of available observing or numerical modeling technology and laboratory or limited-scale field demonstrations.

The other factors that influence community response and consequent decisions are obviously important. For those who will listen, scientific answers about the efficacy of cloud seeding do exist, but science is ever advancing and "final and complete" conclusions are elusive. This aspect often confounds the public, as noted earlier. What would be acceptable as proof of increased water? It is important to have a practical answer to this question. To speculate, statistical terms, for example, scientists may require broadly based evidence of a measurable effect at the 99% significance level, whereas the user or the public may be satisfied with significance at the 80% level (Reinking et al., 1995).

The perception of drought conditions as a predominant decision-influencing factor has stirred some scientific concern. Opponents have often cited cloud seeding as a cause of drought. Also, interest increases when cloud seeding is perceived as an effective drought-relief technology; this is not a surprising finding of the CSU study. However, scientists generally agree that the most soundly based approach to stabilizing water supplies is to conduct sustained rather than crisis-reactive cloud seeding projects. Drought is normally caused by abnormally persistent flow patterns in the global scale atmospheric circulation; this in itself means to scientists that cloud seeding is not a cause of drought. Drought conditions commonly (but not always) offer fewer suitable clouds; seeding in such conditions can be counter to the scientific basis. One cannot expect to make scientists of the populace in a public meeting or two. However, only efforts to educate those concerned can establish a soundly based belief in efficacy and counter the all-too accurate cliché that "interest in cloud seeding is soluble in water" (Farhar and Fitzpatrick, 1990).

2.2.6 Public Participation Procedures

Decision processes that promote public participation create a climate in which the community is satisfied with the decision on whether to go forward with the project. Community members themselves should agree with the decision processes employed. If they believe these processes to be fair, equitable, and responsive to community needs and concerns, they are more likely to accept the decisions resulting from them. Taking the time to receive and respond to public input contributes to a more socially acceptable outcome. Steps that project managers can take to reach this outcome are noted in Table 2-1.

TABLE 2-1. Actions Managers Can Take to Foster Project Acceptance (Reinking et al., 1995)

1. Regard proposed projects as pilots and provide time to learn from implementing them.
2. Involve community stakeholders and local organizations with high credibility.
3. Form citizen review committees, keep them apprised, and listen to them.
4. Convey accurate and complete information to the public, repetitively, through familiar and trusted channels.
5. Explain limitations in knowledge or technique.
6. Identify any potential risks and their magnitude, and develop mitigation strategies.
7. Alert the citizen group and the public to anticipated problems and enlist community members to devise solutions.
8. Listen to feedback, acknowledge it, and modify project design accordingly.
9. Continue work with the community during the project's implementation.
10. Build evaluation components into the project and provide feedback to the citizen group and the public on actual project effects.

2.3 ENVIRONMENTAL ASPECTS

The management of any natural resource, including precipitation, leads to potential ecological changes. Therefore, potential ecosystem changes resulting from increased interactions of precipitation and cloud seeding materials with the environment must be addressed in weather modification projects. Several factors that complicate easy identification and quantification of cloud seeding effects on the environment are discussed here. A brief history of the research on ecological effects of weather modification shows the motivation for pertinent studies. The pertinent concept of cumulative effects is explained (Reinking et al., 1995). Two case studies are examined that deal with potential issues; these exemplify research on ecological impacts of cloud seeding, and environmental impact documentation for a prototype snowpack augmentation program.

2.3.1 Historical Perspective

Several federally funded, multiyear ecology projects have provided vitally needed information on environmental processes that goes beyond the immediate topic of cloud seeding effects on precipitation. The projects included the Medicine Bow Ecology Project (Knight et al., 1975), San Juan Ecology Project (Steinhoff and Ives, 1976), Uinta Ecology Project (Harper, 1981), and Sierra Ecology Project (Berg and Smith, 1980; Berg, 1988). All dealt exclusively with cloud seeding for snowpack enhancement and all were in the western United States. A Project Skywater impact assessment included environmental aspects of increased summertime convective rainfall (USDI, 1977a). Other landmark documents include a technological assessment of winter orographic snowpack augmentation in the Upper Colorado River Basin (Weisbecker, 1974), a more general discussion of ecological effects of weather modification (Cooper and Jolly, 1969), and a comprehensive assessment of potential impacts of artificial ice-nucleating agents (Klein, 1978).

Aside from the Sierra Ecology Project (SEP), the bulk of the research on environmental effects was completed by the early 1980s. The focused investigations addressed concerns such as the direct effects of added snow on large and small mammals and indirect effects through change in type and abundance of food supply, possible physical and biological changes in aquatic systems, and impacts on vegetation. Thereby, the sensitivities of various ecosystems were estimated (e.g., Table 2-2). Some of

TABLE 2-2. Sensitivity Matrix for Selected Environmental Issues (USDI, 1977b; Reinking et al., 1995)

	Environmental setting			
Environmental issue*	Alpine	Forest	Rangeland	Agricultural
Erosion				
Surface erosion	1		1	2
Mass wasting	2	1	2	
Avalanches	2	2		
Microclimate				
Snow duration	2	2		
Soil moisture	1	1	1	1
Water yield	2	2	1	2
Water quality				
Physical	1		1	2
Chemical		1	1	2
Channel processes	1	1	1	2
Sediment yield	2		1	2
Vegetational productivity		2	2	
Nutrient loss	2	2		

*Presumed environmental responses to prolonged precipitation augmentation of 20%. 1 = detectable but insignificant; 2 = readily detectable, sometimes important.

the specific studies and results from the ecology projects are reviewed in the 1983 edition of these "Guidelines" (Committee on Weather Modification of the Irrigation and Drainage Division, 1983).

The essence of the results is that changes that might be expected in the environmental factors (1) were most often subtle, nil, or indiscernible in relation to other natural influences (e.g., effects of fire or insects on forest vegetation); (2) would be of the same type and magnitude as would result from a sustained increase of a corresponding percent in natural precipitation (e.g., as a gradual change in herb species composition might occur in a wetter climate); (3) might be beneficial as often as not and depending on point of view (e.g., as when fish habitat increases with lake level); and (4) would have net outcomes that strongly affect ecosystem management practices (e.g., as when increased weed growth and grassland productivity occur together). During the 1970s, the most common seeding agents, chemical complexes of silver iodide, were examined for ecological effects (Cooper and Jolly, 1970; Klein, 1978). Conclusions from those studies point to little or no effects on terrestrial or aquatic biological communities, either immediately or after many, many years of silver or iodide application in the small dosages possible from cloud seeding (Reinking et al., 1995).

The 1970s ecology studies have credibility, but caution may be appropriately exercised in accepting the conclusions outright. New scientific advances in assessment of ecological response, if applied to cloud seeding, might make earlier results outdated. Even if extreme environmental shifts are not to be expected, potential lesser changes are to be respected. Just as a cultivated crop may respond to enhanced precipitation, so may certain elements in a natural ecosystem. The responses can be manifolding, possibly unexpected, and may follow multiple pathways (Reinking et al., 1995). The past two decades have also seen changing interpretations of, and more rigorous adherence to, regulatory statutes. For example, implications of the National Environmental Policy Act regarding cumulative effects that were completely unaddressed as recently as the early 1980s now often require extensive analysis and documentation. Litigation on environmental issues associated with land management in general has also expanded multifold (Reinking et al., 1995).

The very nature of environmental effects makes them largely site specific. The myriad geologic, pedologic, biologic, hydrologic, and climatic conditions that combine to form the environment of any cloud seeding project area are not necessarily duplicated elsewhere. Although changes in precipitation in the American River Basin of central California may not be expected to induce significant increases in mass wasting, similar changes in precipitation at a more geomorphically sensitive area could be damaging. Equal time must be given to the possibility of positive outcomes; although most research has focused on conceivable environmental impacts in the negative sense, added precipitation or stabilized annual precipitation indeed may be beneficial to the ecology in many instances, except where stabilization or change of any kind in ecology might be regarded as negative (e.g., designated wilderness areas) (Reinking et al., 1995).

Other aspects of the status of the knowledge on environmental effects of cloud seeding directly influence the rigor of projections. These aspects include the research orientation of some of the major ecology projects, and the burden of proof. SEP, for example, was tied directly to the Sierra Cooperative Pilot Project (SCPP). The SCPP was a multiyear research program designed to examine the potential for snowpack enhancement, but not as a long-term operational cloud seeding program. The SEP therefore did not consider the effects of an operational seeding project of unlimited duration and seeding intensity. Also, statistical proof is often lacking in the results of the ecology projects. Early observations emphasized the relatively large variation in both physical inputs (e.g., precipitation amount and timing) and biotic and abiotic responses. The magnitude of the commonly assumed 10% to 15% cloud seeding signal is well within year-to-year variability of natural precipitation [e.g., both record low (462 mm) and high (1704 mm) total annual snowfalls that were recorded at a monitoring station in the central Sierra Nevada occurred within the 6-year duration of the SCPP]. The signal is even more embedded in variations induced by longer-term climatic change. Given that ecosystem responses must react to these wide natural swings, it becomes extremely difficult to rigorously prove anthropogenic cause and environmental effect relationships. Time periods longer than the duration of the typical ecology project are needed to isolate the effects. This phenomenon of low signal-to-noise ratio, in combination with the often intricate and poorly quantified cause-and-effects networks of biotic and abiotic systems, leads to relative ignorance of the timing and magnitude of environmental response (Reinking et al., 1995).

2.3.2 The Concept of Cumulative Effects

Federal regulations (National Environmental Policy Act) define cumulative impact as (Reinking et al., 1995):

> *The impact on the environment which results from the incremental impact of the action when added to other past, present, and reasonably foreseeable future actions regardless of what agency (Federal or non-Federal) or person undertakes such other actions. Cumulative impacts can result from individually minor but collectively significant actions taking place over a period of time.*

The cumulative effects issue has particular significance because cloud seeding typically takes place over extensive geographic areas, and ownership of entire catchments is concentrated among a few individuals, companies, or public agencies. Increasing public and governmental concern make this topic a focal point for formal appeal of proposed management actions.

Several components of the cumulative effects definition warrant discussion. First, the phrasing on the timing of actions (past, present, and reasonably foreseeable future) dictates the need to consider future management activities, many of which are likely to be unknown. Second, since all relevant agencies or persons as potential parties to management actions are included, knowledge of the current and future plans of all parties is required. Private enterprises are not otherwise required to notify the public of their future plans. This often results in a major information void. In situations where both public and private lands are involved, as is typical for cloud seeding operations in the mountainous West, private landholders do not publicize their plans for future land management. Therefore, the "reasonably foreseeable future actions" clause is difficult to implement. Third, the definition forces assessment of combined (collective) effects of individually minor actions (e.g., cloud seeding) with other, potentially more drastic actions. It is conceivable that "minor" effects from cloud seeding become unacceptable in combination with effects from other, completely unconnected actions.

2.3.3 Case Study–The Sierra Ecology Project

The 12-year SEP was designed to study the effects of precipitation augmentation on central Sierra Nevada and Lake Tahoe area snowpack and forest ecosystems due to SCPP. SEP studies generally assumed a 15% maximum annual increase in precipitation in the context of a 5 to 7 year randomized seeding experiment. Results were specific to the target area of SCPP. Background work on the climatologic and hydrologic regimes included evaluations of forest disease and insects (Smith et al., 1978a), deer and their habitat (Smith et al., 1978b), vegetation (Smith et al., 1978c), hydrologic processes (Berg et al., 1980), lake and stream biota (Smith et al., 1980), as well as publication of a bibliography on environmental responses to weather modification (Smith and Berg, 1979). Field studies and monitoring activities then addressed acidity of snowpack runoff, snow-vegetation dynamics, hydrologic consequences of rainfall on the snowpack, and hydrologic disposition of augmented snow cover (MacDonald, 1986).

Two biological communities, subalpine meadow vegetation and mountain hemlock stands, were identified and studied as candidate systems that could change as a result of increased snowpack. These communities were thought to exist in microenvironments that received substantially more water than indicated by regional precipitation amounts. Hemlocks, for instance, occupy sites with cool and moist northern exposures, cold-air drainage, and abundant snow, often immediately below snow cornices. Meadows are often in drainage sumps that concentrate moisture. Among the findings were the following (Reinking et al., 1995):

1. Mountain hemlock stands would increase at the expense of other subalpine vegetation types if snowpack increased over the long term.
2. Some types of hemlock stands would become increasingly less species rich and the size-structure patterns of hemlocks would be altered.
3. Snow cover duration was a principal control of subalpine meadow plant development, reproductive success, and existing vegetation patterns.
4. Long-term effects of snowpack increased by 30% every year might result in dramatic vegetation changes, but longer-term studies would be needed to substantiate this assertion.

The hydrologic effect of an extended snow cover was investigated in a field experiment by sprinkling water on 1,000 square meter plots in a small basin for 11 days after natural snowmelt ended (MacDonald, 1986). Rises in groundwater were observed, but once the simulated snowmelt ceased, piezometric pressures appeared to decline more rapidly than before. Other data supported the contention that increases in streamflow resulting from snowpack augmentation would not extend appreciably beyond the time of soil desaturation.

Although the aim of snowpack augmentation projects is not to enhance rainfall, potential impacts from increased rainfall stood out in the SEP as the single most critical concern. In uncommon circumstances (e.g., increased rainfall on steep, disturbed, saturated soils), an increase in soil erosion by splash, sheet, or rill erosion is possible. Similarly, if snowpack augmentation resulted in a thin snow cover on ground that would otherwise have been bare, snowmelt during subsequent rainfall would lead to greater runoff than would have occurred naturally. Although these concerns are real ones, routine adherence to suspension criteria that guard against rain-induced flooding would reduce the likelihood of appreciable negative impacts (Reinking et al., 1995).

SEP recommendations included the monitoring of environmental situations anticipated to concentrate the effects of snowpack increases. Subalpine meadows and mountain hemlock stands were identified as candidate biotic communities. Snowbanks and spring/seep communities were also candidate index sites for long-term monitoring.

2.3.4 Case Study–Environmental Impact Statement for a Prototype Project

The Lake Oroville Runoff Enhancement Program was a 5-year prototype project under way in the Feather River drainage of California's Sierra Nevada. It was designed to augment snowpack by cloud seeding from ground-based generators with liquid propane as the seeding agent. The

Plumas National Forest and the California Department of Water Resources (CDWR), the lead public agencies responsible for the proposed project (USDA Forest Service and CDWR, 1990; USDA, 1991), completed a comprehensive joint Environmental Impact Statement/Environmental Impact Report (EIS/EIR) in September 1990. The document was written to allow tiering to similar projects (should they be proposed). A wide range of potential environmental issues was assessed. These included water resources (e.g., rain-snow level, length of winter, snowpack, extent of delayed snowmelt, groundwater, avalanches, runoff and floods, water use, and downwind precipitation depletion), erosion, water quality, plant communities, rare plants, wildlife, fish and aquatic life, endangered and threatened animals, cultural resources, aesthetic values, transportation, and safety (e.g., floods and avalanches, hazardous material spills, and fire hazard). Within these relatively broad categories, a wide variety of specific sub-issues were identified, such as those associated with the installation and use of the propane ice-nucleating generators.

Some results from the joint EIS/EIR elucidate the extent of the investigation (Reinking et al., 1995):

1. Installation of temporary facilities (the propane generators) on land allocated to semi-primitive management by the Forest Service will be removed in summer, painted white to reduce aesthetic impact, and located away from the hiking trails.
2. Land-disturbing activities (e.g., the propane dispensers and precipitation gauges) can be minimized by the careful replacement of the disturbed soil as near to original conditions as possible to reduce the possibility for increased erosion.
3. Changes in the amount and intensity of the snowpack and rainfall are expected to be well within natural variations.
4. The contribution of non-combustive propane seeding to the "greenhouse effect" is negligible in the anticipated seeding environment.
5. A delay in snowmelt of 0 to 3 days is estimated.
6. The propane seeding of cold winter storms will not have a depletable effect on precipitation downwind of the seeding area.
7. There should be no measurable direct effect on erosion from an augmented snowpack within the project area.
8. Plant species with extremely limited habitats, including narrow tolerance to soil moisture regimes, may be affected. Since precipitation would not be altered from the normal range, sensitive plant populations within the project area are not expected to change.
9. Cultural resources (e.g., prehistoric petroglyphs and historic buildings) have weathered the natural, extreme ranges of measured snowfall in the area and will not be affected by the probable increase in snowfall.

The joint EIS/EIR was appealed to the Regional Forester of the Forest Service's Pacific Southwest Region on numerous issues, many dealing with a lack of site specificity and inadequate analysis of direct, indirect, and cumulative impacts of the proposed project. In response to the appeal, the Regional Forester affirmed six issues raised by the appellants. Those issues were:

1. The EIS/EIR did not adequately describe the existing known data that can relate to the watershed condition and fisheries habitat of the third-order streams.
2. There was not an adequate description of the cumulative effects and the factors used in the cumulative watershed effects analysis on the third-order drainages.
3. The effects of the project on sensitive, threatened, and endangered wildlife species need to be better addressed.
4. A further analysis needs to be made on the potential effects of flooding on small streams.
5. Identify if there are any municipal supply watersheds within the project area, and, if so, the effects of the project on water quality in these watersheds.
6. Assure that the California Department of Fish and Game and the U.S. Fish and Wildlife Service are consulted on this project.

A supplement to the EIS/EIR for the project summarized the original EIS/EIR and successfully addressed these issues (USDA, 1991). In all, several stages of development, response to identified issues, and levels of approval were required in the EIS/EIR process. A negative (no effect) declaration (which was not accepted), the original EIS/EIR (which was appealed), and the supplement EIS/EIR (which received final approval from the Forest Service in November 1991) each required approximately one year to prepare in succession, at a total cost of $400,000 or more. At an additional cost of some $100,000 annually, continuous environmental monitoring of watershed effects was put in place for the prototype cloud seeding project.

2.4 ECONOMIC ASPECTS

It is unlikely that every effect of added rain or snow will be anticipated, but efforts should be made to predict as many changes and impacts as possible (Stroup, 1973). Rational decision making still calls for a valid benefit-cost analysis. Evaluating economic benefits and liabilities of weather modification accurately requires an understanding of the technology itself. In all, the practitioner has much better information than existed a decade ago, but the following general assessment, without regards to an indexed year for the reported dollar amounts, still remains in force.

2.4.1 Deciding the Goal and Scale of Economic Analysis

Many studies highlight the complexities of identifying and weighing anticipated economic benefits and costs to various segments of society and elements of the economy. "Failure to distinguish between net economic benefit and transfer effects (as in redistribution of wealth) has given rise to endless difficulty in the analysis of water projects in the past," Crutchfield (1973) advises. "There is no need to perpetuate the confusion in assessing new technologies." Appropriate measures of desirable or undesirable financial effects of enhanced precipitation are possible but depend on point of view, whether local, national, or global, and the geographical extent and nature of economic impact is a critical issue.

Uncertainties in the technology add to uncertainties in the economic evaluations, but to some degree this is true in any area of interest. A global view of aspects appropriate to consider is stated by Sonka (1979):

> *Because weather events can have severe adverse effects on economic activity, the gross benefits of successful weather modification activities are apparently very high. And, in general, the operational costs of modification activities are small relative to those gross benefits. (But) the existence of such positive net benefits does not insure that some individuals would not suffer substantial decreases in welfare. Probably the most important aspect in determining the credibility of any economic analysis, however, is the viewpoint of that analysis. ... It should be clear that the goal of the analysis is to determine the effects of the modification activity on the entire economy of a region, not just impacts on those sectors which derive benefits from the planned activity.*

This view of a comprehensive economic evaluation must be tempered by the scope of an operational project and the practical capabilities of the organization or enterprise conducting or proposing to conduct the operations. This view does not imply that private enterprise or other individual or collective sponsoring organizations have no role in providing cloud seeding services. It simply means that a thorough economic evaluation of cloud seeding must involve calculated socioeconomic costs and benefits much more comprehensive than those that may be appropriate concerns for the private operator or sponsoring organization (Crutchfield, 1969; Reinking et al., 1995).

Economic analyses for the adoption of cloud seeding technology in specific situations do offer guidance not only for vested stakeholders:

> In view of the uncertainties that attend the application of weather modification by cloud seeding, one might inquire why it has been so

widely adopted around the world. The answer, of course, lies in the large economic benefits it offers. Although difficult to measure exactly, the perceived benefits far outweigh the cost of implementing programs. Although a certain economic objective might be attainable through cloud seeding, this is not in itself sufficient justification for launching an operational program (Dennis, 1980).

In all, there are no simple answers or alternatives in assessing costs and benefits of cloud seeding. There are, instead, many gray areas of compromise that must be acknowledged and considered (Dennis, 1980). This is illustrated in the following discussion, where the economics for two different situations are considered: when cloud seeding is applied to enhance direct rainfall on crops (generally summer cloud seeding), and when additional water from seeding is impounded and released in a controlled manner for later use (generally winter cloud seeding) (Reinking et al., 1995).

2.4.2 Economic Aspects of Summer Cloud Seeding

A number of studies of economic effects of summer precipitation enhancement have been conducted in the United States, primarily for the High Plains and the Midwest. The approaches and results reveal appropriate response variables and the levels of sophistication possible in such assessments as well as fundamental differences dictated by geography (Reinking et al., 1995).

2.4.2.1 High Plains. The economic effects of added growing season rainfall on High Plains agriculture were examined at the North Dakota Agricultural Experiment Station (Schaffner et al., 1983). Five objectives were specified:

1. Measure dollar values of direct benefits to farmers and ranchers of added growing season rainfall.
2. Determine enterprise adjustments needed on the farm and ranch to respond most profitably to increased rainfall.
3. Measure the total added direct benefits to the four farming areas and the state.
4. Examine the broader enterprise shifts within farming areas due to added growing season rainfall.
5. Measure the total impact of added growing season rainfall on the economy of the state.

Depending on the farming area, added rainfalls of 2 cm for June to July and 3 cm for June to August were used to determine crop, livestock, and economic responses. The rain increases would be realized over many consecutive growing seasons. The added quantities represent much larger percentage changes for the western than the eastern parts of the state, because the west is normally much drier. Linear programming models were applied to select the most profitable farm enterprise plans with normal and added rainfall. Changes expected in yield from 15 different crops and forages and in livestock numbers were estimated. Five-year average prices for crop and livestock products in each farming area were used in the economic analysis. An input/output model based on actual expenditure of North Dakota businesses was used to account for spending and respending to estimate responses of other sectors of the economy (e.g., finance, wholesale and agricultural processing, construction, retail, real estate, services, households, government) (Reinking et al., 1995).

In 1977–1981 average dollars, the assumed increases in rainfall produced increases in direct returns over variable costs to agriculture for the western, west central, east central, and Red River Valley (eastern) farming regions of, respectively in millions of dollars, \$53.0 (19%), \$48.3 (14%), \$47.1 (10%), and \$29.2 (7%). To meet these economic enhancements, changes in relative acreages occurred in each area to accommodate the most profitable crops and forage. Livestock production increased significantly in all areas, but progressively more from east to west (Reinking et al., 1995).

Direct purchases by farmers from the local economy resulting from increased farm income and expenditures totaled \$226 million for the state in the model estimates. This total when spent and respent generated another \$450.5 million of estimated gross business volume, and the largest increases occurred in the driest (western) part of the state (Table 2-3). Business volume in this simulation increased in all economic sectors, and the largest increases were associated with households and retail (Reinking et al., 1995).

2.4.2.2 Midwestern United States. Rainfall variability and differential geographical effects on selected agricultural areas were considered in a study of the potential benefits of cloud seeding in Kansas (Smith, 1978). An assumed precipitation alteration scheme that varied changes in rainfall rate and amount from a 10% decrease to a 75% increase was applied to a 30-year series of rainfall observations. The simulated average growing season rainfall increased in a range from 3.8 cm in southeastern Kansas to 5.7 cm in northwestern Kansas. The expected changes in yield (Table 2-4) illustrate that not all crops respond proportionately, and the total response depends on the climatic and agricultural conditions of the region. Benefits of added grain crop production were linked to the price conditions assumed. In the western region, where benefits were potentially the largest, the 1978 estimates ranged from \$99 million to \$127 million (Smith, 1978). The higher estimate assumed no reduction in crop prices from increased production; the lower estimate considered such a reduction. With lower prices, the study noted, producers in areas not directly affected by cloud seeding might experience loss of income, or the western part of Kansas might gain at the expense of eastern Kansas if a successful rain enhancement program were instituted statewide (Reinking et al., 1995).

TABLE 2-3. Estimated Increased Business Volume Resulting from Increased Farm Income and Expenditures in U.S. Dollars (millions) Generated by Added Growing Season Rainfall by Farming Areas in North Dakota (Schaffner et al., 1983; Reinking et al., 1995)

Economic Sector Receiving Added Business	Western	West Central	East Central	Red River Valley
Added Farm Income and Expenditures	67.3	59.6	58.6	40.5
Economic Sector to Which Business Volume Accrued:				
Agricultural livestock production	5.3	4.3	4.4	3.0
Agricultural crop production	3.3	1.7	2.2	1.3
Sand and gravel	0.0	0.3	0.3	0.2
Construction	5.7	5.0	4.8	3.1
Transportation	1.1	0.8	0.7	0.5
Communication and utilities	6.8	6.0	5.8	3.9
Wholesale and agricultural processing	6.7	2.6	4.2	2.1
Retail	51.9	46.1	47.1	33.8
Finance, insurance, and real estate	12.3	10.2	9.7	6.8
Business and personal services	5.3	7.0	3.8	3.9
Professional and social services	8.0	6.5	6.4	3.9
Households	93.0	81.6	79.8	51.1
Government	6.8	5.9	5.7	3.8
Total Added Business Volume	206.2	178.0	174.9	117.4

TABLE 2-4. Average Expected Yield Changes Due to Assumed Precipitation Alteration in Kansas. (Smith, 1978; Reinking et al., 1995)

Crop	Eastern Kansas	Central Kansas	Western Kansas
Fallow wheat	—	+0.46 bu/acre	+4.76 bu/acre
Grain sorghum	+0.47 bu/acre	+0.25 bu/acre	+3.15 bu/acre
Continuous wheat	−0.07 bu/acre	+0.88 bu/acre	+2.31 bu/acre
Forage sorghum	—	+0.18 ton/acre	+0.17 ton/acre
Soybeans	+0.36 bu/acre	—	—
Alfalfa	+0.09 ton/acre	—	—
Corn	−0.02 bu/acre	—	—

(bu/—bushels per)

In a related unique field experiment in Illinois, the actual effects of enhanced rain on crop production were evaluated. Large (9 m $\times$ 48 m) mobile, plastic-covered shelters with sprinkler systems were used to exclude natural rain but otherwise expose crop plots to the prevailing weather. Watering is quantified and timed to the historical rain-day precipitation records for wet, dry, and average summers, and water is added to simulate modification. Initial results indicated that rainfall increases of 10% to 40% in Illinois increase corn and soybean yields by 4% to 20% if natural rainfall is below or near average (Changnon and Hollinger, 1988).

Refined results showed that, for 2.5 cm of rainfall added during a hot, dry summer, yields increased 10 bushels per acre for corn and 4 bushels per acre for soybeans. In a summer of average rain, increases are less, about 5 bushels per acre for corn and 3 bushels per acre for soybeans. Yields of both crops were shown to decrease when summer rainfall exceeds 36 cm. Rain increases of realistic percentages applied with the sprinklers only on days when natural rainfall was less than 0.25 cm provided no detectable yield increases, whereas a 40% rain increase on all rain days produced the greatest increase in crop yield. Corn yields responded well to added rain on days with 0.25 to 2.5 cm of natural rainfall (Changnon and Hollinger, 1990).

Yield trends and stability influence both microeconomic (e.g., single farm) and macroeconomic (e.g., aggregate Corn Belt) decision makers as they set priorities for investment in new technologies, such as cloud seeding (Garcia et al., 1987). It is not feasible for a single producer to have a cloud seeding program, but groups of producers might do so. Thus, it is appropriate to consider local and regional impacts (Dennis, 1980), i.e., the level of aggregation in effort, area seeded, and economic effect (Reinking et al., 1995).

As yields have steadily increased with improving agricultural technology other than cloud seeding, variability of yields and the absolute yield risk to producers have also increased. These increases in variability and risk could be due to a heightened sensitivity of technology to weather, or to temporal increases in weather variability. These findings have implications for estimating the economic effects of a fluctuating climate and society's response to mitigate adverse effects. Enhanced precipitation and consequent moderated crop heat stress might reduce risk by alleviating extreme year-to-year yield changes. However, yields are influenced by a broad set of agriclimatic conditions that must be considered in estimating the overall impact of weather. Therefore, differences in estimated weather effects on yields for similar aggregation levels must be accounted for. Lack of sensitivity to these differences can result in inappropriate measurement of the distribution of economic gains from activities such as cloud seeding. The effects of seeding are likely to be regionalized, and this can distort its relative attractiveness to user groups representing differing spatial aggregations such as a few farms, a few counties, or a state (Garcia et al., 1987).

To aid in policy making, an effort was made to determine the economic beneficiaries of a functioning precipitation modification technology when applied at various spatial and temporal aggregations (Garcia et al., 1990). The results of the simulation illustrate potential distributions of economic effects, demonstrate the importance of careful planning in the use of weather modification technology, and provide information that is useful in determining the roles of local, state, and federal governments in support of weather modification (Reinking et al., 1995). Differences in soil types, climatic conditions, and crop response all influence producer revenues for a region of given size with precipitation enhancement over a given time period. According to the Garcia et al. (1990) study, producers within small target regions (with increased precipitation and crop yields) realize the largest revenue gains. Producers outside the small target experience only small revenue reductions due to the increased competition. The added revenue from a small area within a larger target region declines, and the revenues of producers in adjacent non-target regions are reduced much more, as the size of the total target region increases. The econometric simulation led to the conclusion that "for programs covering multistate areas, the change in total (producer) revenues to the target areas is negative." However, in the simulation, consumer savings increase with the size of the target area of successful precipitation enhancement. These results stem from changes in the regional and national balances of supply and demand. An increase in revenue for producers from added precipitation over time favors multiyear use of precipitation modification technology over one year isolated use. However, it is cautioned that successful precipitation enhancement continued over several years may change production and technological responses within and outside the target areas, and change consumer and producer benefits (Garcia et al., 1990).

2.4.3 Economic Aspects of Winter Cloud Seeding

Winter cloud seeding to augment snowfall in high-elevation areas is designed primarily to increase runoff for hydroelectricity and water supplies for lower elevation, semiarid areas (Foehner, 1983). In this situation, the beneficiaries usually do not reside in the project area. Projects conducted to enhance winter sports activities are an exception. In either situation, the economic value of additional water can be calculated somewhat more readily than in cases in which crop response is directly involved (Reinking et al., 1995). Estimated benefit/cost (B/C) ratios of 23.5:1 to 9.7:1, respectively, for enhanced runoff for hydroelectric power production were reported by Griffith and Solak (1999, 2002).

Managed water is normally assigned a value equal to the cost of obtaining it, storing it, transporting it to the region of use, and distributing it to the users (Dennis, 1980). However, water and electricity supply and demand also influence its value. Favorable effects on the net economic productivity of a hydroelectric utility system include more efficient use of storage capacity, a favorable change in the ratio of peak to average plant capacity, and a reduction in the overall capital intensity of the hydro-generating system (Crutchfield, 1969). Snowpack managed for the winter sports industry is used where it falls, and it directly benefits the many industries associated with skiing. The availability, and direct benefits and costs, of the managed additional snowpack or runoff will have ripple effects in other economic sectors. One might think that factors affecting the value of additional water would be the same throughout the western United States. However, there is actually considerable diversity in the factors that determine the cost and value of (added) water (Reinking et al., 1995).

2.4.3.1 Arizona and Nevada. Arizona receives some 90% of its renewable water supply from winter precipitation. This state relies on groundwater for 40% of its water supply, even with the newly opened aqueduct provided by the Central Arizona Project. According to the Arizona Department of Water Resources, overdraft of aquifers can be 2.5 million acre-feet[6]/year; this can lower the groundwater level by several hundred feet, making it cost-prohibitive to pump, causing land subsidence, and introducing water quality problems. New considerations are being given to water augmentation and reuse programs; indeed, Arizona public policy, driven by economic and environmental considerations, officially acknowledges cloud seeding as a possible water augmentation method (Reinking, 1992; Gelt, 1992). The Salt River and Verde River watersheds contribute 1 million acre-feet/year; here alone, a 15% increase in runoff would meet the needs of 750,000 people annually. For Nevada, the University of Nevada Desert Research Institute indicates that the 1990 valuation of urban water rights is about $2,500/acre-foot, and runoff yields from precipitation enhancement at 0.025 cm/hr for normal hours of precipitation would cost about $10/acre-foot (Reinking et al., 1995).

2.4.3.2 Utah. Utah is the second-driest state in the United States. The winter snowpack and associated runoff are necessary for agricultural and

[6]acre-foot = 1.23335×10^3 m^3. Acre-foot or acre-feet, where used in this document, is quoted in the referenced sources, and may be most useful to the intended audience.

urban supplies and for the ski industry. The state spends some $8.5 million annually (1990 valuation) on water development, according to the Utah Division of Water Resources (UDWR). Demands for urban use have increased with the state's population, an increase of about one-third since 1990. As in other states, early season snowfall is highly valued by the ski industry. The agricultural need is for late-season irrigation water which is valued near $40/acre-foot, according to the UDWR, whereas the estimated direct cost of water from an 8% to 12% increase in snowpack from cloud seeding in key mountain watersheds is $10/acre-foot. In recent years, as many as 18 counties or water conservancy districts have contributed nearly $0.5 million to cost-share operational cloud seeding with the state. This level of cost-sharing has continued, and has been considered a reasonable benefit-cost risk. The state has continued to collaborate with the federal government on research to determine the actual efficacy of cloud seeding (Reinking and Meitin, 1989; Reinking, 1992).

Benefit-cost ratios of 3:1 to 10:1 were estimated for 10% mountain snowfall increases in the Sevier River basin in Utah (Super and Reynolds, 1991). The basis was the amount of additional water potentially produced, the estimated value of the additional stream flow, and the direct cost of conducting an effective operational program. Not accounted for were other benefits and costs that would affect an aggregate benefit-to-cost ratio. The 10:1 ratio reflects costs of the Utah operational seeding program as currently conducted, using a mixture of manually operated valley, canyon-mouth, foothill, and low-mountain generators well up the windward slopes to improve targeting. The National Weather Service River Forecast Center in Salt Lake City estimated that about 10,000 acre-feet (14%) would be produced in the mean annual runoff by the assumed 10% snowfall increase above the 2,440 m elevation (Figure 2-2 of Reinking et al., 1995); this estimate was derived with a snow accumulation and ablation model that numerically accounts for various physical processes taking place in the snowpack (Anderson, 1973). The 3:1 ratio reflects the additional costs of using closely spaced, high-operated valley, canyon-mouth, foothill, and low-mountain generators. The UDWR then estimated the value to the land areas along the Sevier River drainage at about $18/acre-foot or $180,000 annually. For the 880-square-kilometer high-elevation target, $10,000 was the annual cost of the standard operation; $53,000 was the cost for the season with the addition of the high-altitude seeding option (15 additional remote-controlled generators, capital costs, and seeding materials) (Reinking et al., 1995). In a more recent UDWR study (Stauffer and Williams, 2000) and in Stauffer (2001), it was estimated that the operational winter cloud seeding projects in Utah generate approximately 250,000 acre feet of additional runoff. These UDWR studies estimate the cost of producing the additional water to be approximately \$1/acre-foot via the current operational winter orographic cloud seeding program methodology used throughout Utah.

2.4.3.3 California. Because precipitation in California varies extremely with latitude, and surface water is transported from the wetter north to the drier south, water values are typically much greater in the south. Economic analyses commonly segregate water value by use, with hydroelectric power generation, agriculture, and in-stream uses listed most often. The geographic distribution of each of these uses, in addition to the dryness of the year, partially determines water value (Reinking et al., 1995).

In 1992, a dry year, the California Department of Water Resources (CDWR) judged the average value of "new" water from the Feather River basin of northern California to be \$30/acre-foot. The CDWR Drought Water Bank, in 1992, paid \$50/acre-foot for real, new water at the Delta in the Sacramento Valley and marketed it for \$70 to \$75/acre-foot in the same location after accounting for conveyance losses and other charges. In 1991, also critically dry with larger urban shortages, the Water Bank paid \$125/acre-foot which translated into a price of \$175/acre-foot plus transportation costs to users south of the Delta. Water values are at least \$150/acre-foot in the south coastal area, and even higher in some local areas in the south. One analysis listed total agricultural and hydrogeneration value of water for the Kings River (southwest slope of the Sierra Nevada) as more than \$320/acre-foot in 1986 (Romm and Ewing, 1987). Henderson (2003) reported benefit/cost ratios of six long-running programs in California. Benefit/cost ratios were calculated to range from 13:1 to 61:1, based on precipitation increases of 2% to 9%. The study valued the additional water at \$60/acre-foot and additional electrical generation at \$20/Mwh, both conservative estimates. Opportunities to generate additional water by cloud seeding are generally fewer in the south, but the higher value may make projects with lower yield worthwhile (Reinking et al., 1995).

2.4.3.4 Colorado. Approximately 70% of Colorado's water is supplied by snowmelt runoff (Sherretz and Loehr, 1983), and snow is extremely valuable while on the mountains because winter sports have surpassed agriculture as the state's leading industry. Six major runoff-producing areas within the Colorado River Basin have a total high-water yield area of 58,500 square kilometers. If cloud seeding could produce 1.43×10^6 acre-feet annually within the Upper Basin (approximately 10% of the average annual streamflow), and an additional 0.83×10^6 acre-foot in the lower and adjacent basins, of the total, "approximately . . . 1.7×10^6 acre-feet would be available to reduce deficits and meet new demands. Valuing this water at . . . \$30/acre-foot, the total benefit from additional water would be \$48.5 million/year" (Lease, 1985). This estimate is based on a computer simulation of the impact of additional runoff produced by cloud seeding. The model of the Colorado River reflects water availability, salinity, and demands on water by municipal, industrial, energy, agricultural, and other users. Based on projected time and water demand relationships made for points along the river, the impact on river water supply and quality can be predicted (Lease, 1985).

The possible increases in streamflow from cloud seeding could significantly increase the quantity and value of energy output from small-scale hydropower facilities in Colorado (Loehr et al., 1983). Given a value on electric power, Loehr et al. developed a method for evaluating the impact of weather modification from a run-of-river facility and a conventional dam. They found the wholesale value of power from such facilities ranged from $0.014 to $0.12/kilowatt-hour, depending on the circumstances in which the energy is produced and used. For two sites studied, they estimated that a 15% increase in the April 30-snow water equivalent increases electric energy output by 3.5% to 6.1% and its total value by 5% to 9.9% annually (Reinking et al., 1995).

For the Colorado ski industry, any delay in opening-day/early-season snowfall, or slow business at Christmas due to lack of snow, substantially affects the state's economy. Sherretz (1983) statistically estimated that 15% snowfall increases for hypothetical dry winters at Colorado ski areas are associated with 2% to 8% increases in total season visits (skier visits equal the number of lift tickets sold). His "conservative" estimates of retail expenditures by these additional skiers were in the $0.5 to $10 million range for six ski areas, during the early 1980s. The activity in the Colorado ski industry has magnified many fold since then, so early-season snow is all the more important to state economics. The antithesis is that "additional snow in the midseason probably does not significantly increase skier visits and attendant retail expenditures" (Sherretz, 1984). The Colorado counties that do not host ski areas benefit from wages paid to residents who commute to jobs at ski areas (Reinking et al., 1995).

Snowmaking machines at the lower elevations of the ski slopes have greatly increased the stability of opening day and low-snow periods. However, some ski areas do employ cloud seeding, especially to try to ensure reliable snow cover at the highest elevations. The relative value of the benefits might be estimated by determining the cost of meeting the requirements by these alternative methods (Reinking et al., 1995).

A frequently expressed concern of residents in cloud seeding project areas is that more snow will mean greater snow-removal costs. From a local viewpoint, it has been suggested that citizens whose lifestyles and incomes are negatively impacted by enhanced precipitation may require compensation, which would be computed on variables such as wages lost or costs incurred from incrementally more adverse weather (Weisbecker, 1974). However, it has been found to be very difficult to assess the cost of removing an additional increment of snow (Reinking et al., 1995).

Responding to such concerns, the Colorado Department of Natural Resources assessed county snow-removal procedures and developed a computer model to simulate snow-removal costs (Sherretz and Loehr, 1983). Costs were simulated because most counties do not keep detailed records of snow-removal expenses. Information on wages, equipment, and removal procedures was obtained by interviewing road maintenance foremen. Variations in these factors and snow-removal strategies are reflected in the time required to remove a certain amount of snow in different counties (Reinking et al., 1995). A large increase in snowfall (25%) due to cloud seeding was assumed in one-third of the observed storms. Estimated costs per employee for removing snow from unseeded storms ranged from $1,300 to $11,000 in a winter of heavy snowfall. Additional snow from seeding was estimated to increase removal costs by 0.8% to 12.6% in a heavy snowfall winter. The average cost increase over all counties studied was 6.1% in winters with heavy and average snowfall and 4.9% in winters with low snowfall. Somewhat in contrast, the California Department of Transportation (unpublished) found that the increases in snowfall in near average or below average precipitation years are within the range covered by the major fixed costs for equipment and labor. In California, spring flooding can be a problem, so cloud seeding operations might be suspended in years when snow depths significantly exceed average. Colorado Mountain and Western Slope counties were encouraged to develop procedures for collecting data of sufficient detail to compute snow-removal costs accurately. Such aspects are aggregate effects, similar to those identified with summer seeding (Reinking et al., 1995).

2.5 CONCLUSIONS

Social, environmental, and economic factors will determine whether or not a cloud seeding program is accepted. Risk-benefit assessments are appropriate for each of these categories. Many levels of sophistication are possible in measuring, analyzing, and projecting the intertwined socio-environmental-economic effects. Benefits of cloud seeding are determinable for increased hydropower generation, salinity reduction, enhanced snowpack for ski industries, and increased water supplies for fish and wildlife, recreation, municipal, industrial, and agricultural users. Also, each of these potential benefits carries some liabilities. The possibilities and costs of meeting the requirements by alternative methods, and the consequences of having no such technology, should be considered. Appropriate measures of possible direct and indirect effects, and divisions of responsibilities for providing the analyses and dealing with their outcomes, must be determined.

The potential sphere of influence of the cloud seeding operation and the goals of the analyses are important considerations. The heterogeneity of the affected population gives rise to diverse goals and the potential for controversy. Awareness of public concerns, a responsive and well-guided public information/involvement program, a corresponding decision process, and ongoing evaluation of both the direct and indirect effects will provide many appropriate checks and balances as a cloud seeding program is brought from concept to application. The fundamental principles to be applied to successful siting and operation are not unique to cloud seeding projects. The same kinds of issues are encountered with power transmission lines, waste-to-energy conversion facilities, and nuclear-waste disposal facilities, and a multitude of other less "newsworthy" endeavors that may benefit mankind if properly guided and managed. Whether to temper a view of the risks of conducting cloud seeding with a perspective for the risks of not understanding and developing alternative technologies for providing adequate water supplies is a matter of social choice. The startup and continuation of a cloud seeding program will most likely depend on the perceptions of the benefits and liabilities as derived from this whole process, the diligence with which effects are monitored, and how well public involvement is maintained (Reinking et al., 1995).

2.6 REFERENCES

Most of these references were carried over from Reinking et al. (1995). However, many figures provided in the previous edition have only been referenced by a general reference to that edition and only the process within some figures has been described in this edition.

American Society of Civil Engineers (2003). *Standard Practice for the Design and Operation of Hail Suppression Projects.* ASCE/EWRI Standard 39-03, Reston, VA., 59 p.

ASCE (2004). *Standard Practice for the Design and Operation of Precipitation Enhancement Projects.* ASCE/EWRI Standard 42-04, Reston, VA, 63p.

ASCE (2005). *Standard Practice for the Design and Operation of Super-Cooled Fog Dispersal Projects.* ASCE/EWRI Standard 44-05, Reston, VA, 30p.

Anderson, E. (1973). "National Weather Service River Forecast System Snow Accumulation and Ablation Model." *NOAA Technical Memo NWS HYDRO-17,* U.S. Department of Commerce, Silver Spring, MD.

Berg, N.H. (1988). "A Twelve-Year Study of Environmental Aspects of Weather Modification in the Central Sierra Nevada and Carson Range." *The Sierra Ecology Project,* Unpublished report on file at the Pacific Southwest Research Station, Forest Service, United States Department of Agriculture (USDA), Albany, CA.

Berg, N.H., and Smith, J.L. (1980). "An Overview of Societal and Environmental Responses to Weather Modification." *The Sierra Ecology Project,* **5**, Office of Atmospheric Resources Management, Bureau of Reclamation, U.S. Department of the Interior (USDI), Denver, CO.

Berg, N.H., Bradford, W.L., Brown, K.J., Menke, J.W., Singer, M.S., and Smith, J.L. (1980). "An Evaluation of Possible Effects of Weather Modification upon Hydrologic Processes in the American River Basin, California." *The Sierra Ecology Project*, **4**, (2), Office of Atmospheric Resources Management, Bureau of Reclamation, USDI, Denver, CO.

Borland, S.W. (1977). "Hail Suppression: Progress in Assessing its Costs and Benefits," in *Hail: A Review of Hail Science and Hail Suppression*, G. B. Foote and C. A. Knight, (eds.), *Meteorological Monographs*, **16**, (38), American Meteorological Society (AMS), Boston, MA, 155–175.

Changnon, S.A., Jr., and Hollinger, S.E. (1988). "Use of Unique Field Facilities to Simulate Effects of Enhanced Rainfall on Crop Production," *J. Weather Modification*, **20**, Weather Modification Association (WMA), Fresno, CA, 60–66.

Changnon, S.A., Jr., and Hollinger, S.E. (1990). "Crop Yield Results from Simulated Rain Applications to Agricultural Plots in Illinois." *J. Weather Modification*, **22**, WMA, Fresno, CA, 58–62.

Changnon, S.A., Jr., and Lambright, W.H. (1990). "Experimentation Involving Controversial Scientific and Technological Issues: Weather Modification as a Case Illustration." *Bull. Amer. Meteor. Soc.*, **71**, AMS, Boston, MA, 334–344.

Changnon, S.A., Jr., Davis, R.J., Farhar, B.C., Haas, J.W., Ivens, J.L., Jones, M.W., Klein, D.A., Mann, D., Morgan, G.M., Jr., Sonka, S.T., Swanson, E.R., Taylor, C.R., and Van Blokland, J. (1977). "Hail Suppression: Impacts and Issues." *Final Report, Technology Assessment of the Suppression of Hail*, Illinois State Water Survey, Urbana, IL.

Committee on Weather Modification of the Irrigation and Drainage Division (1983). "Guidelines for Cloud Seeding to Augment Precipitation." *J. Irrigation Drainage Engineering*, **109**, (1), ASCE, New York, NY, 111–182.

Cooper, C.F., and Jolly, W.C. (1969). "Ecological Effects of Weather Modification: A Problem Analysis." *Department of Resource Planning and Conservation*, University of Michigan, Ann Arbor, MI.

Cooper, C.F., and Jolly, W.C. (1970). "Ecological Effects of Silver Iodide and Other Weather Modification Agents: A Review." *Water Resources Research*, **6**, American Geophysical Union, 88–98.

Crutchfield, J.A. (1969). "Economic Evaluation of Weather Modification," in *Weather Modification, Science and Public Policy*, R. G. Fleagle, (ed.), University of Washington Press, Seattle, WA.

Crutchfield, J. A. (1973). "Social Choice and Weather Modification: Concepts and Measurement of Impact," in Modifying the Weather, A Social Assessment W. R. D. Sewell et al. (eds.), *Western Geographical Series*, **9**, Department of Geography, University of Victoria, Victoria, B.C., Canada, 187–228.

Dennis, A.S. (1980). "Weather Modification by Cloud Seeding." *International Geophysics Series*, **24**, Academic Press, New York, NY.

Farhar, B.C. (1973). "The Relationship of Socio-Demographic Characteristics and Attitudes Toward Weather Modification: The South Dakota Study, 1972." *J. Weather Modification*, **5**, WMA, Fresno, CA, 261–276.

Farhar, B.C. (1974). "The Impact of the Rapid City Flood on Public Opinion about Weather Modification." *Bull. Amer. Meteor. Soc.*, **55**, AMS, Boston, MA, 755–758.

Farhar, B.C. (1975a). "Weather Modification in the United States: A Socio-Politico Analysis," *Ph.D. Dissertation*, Department of Sociology, University of Colorado, Boulder, CO.

Farhar, B.C. (1975b). "Weather Modification and Public Opinion in South Dakota, 1972 and 1973." *J. Appl. Meteorology*, **14**, AMS, Boston, MA, 702–709.

Farhar, B.C. (1976). "Hail as Sudden Disaster: Public Response to Hail Suppression Activity." Mass Emergencies, Special Issue: *Environmental Stress: Threat and Response in Various Social Systems*, **1**, 313–321.

Farhar, B.C. (1977). "The Public Decides about Weather Modification." *Environment Behavior*, **9**, 279–309.

Farhar, B.C. (1978). "What Does Weather Modification Need?: A Societal View." *J. Appl. Meteorology*, **17**, AMS, Boston, MA, 878–888.

Farhar, B.C., and Clark, J. (1978). "Can We Modify the Weather?" *A Survey of Scientists*. Institute of Behavioral Science, University of Colorado, Boulder, CO.

Farhar, B.C., and Fitzpatrick, C. (1990). "Weather Modification in Winter and Summer: A Comparative Study of Community Response." *Final Report, National Science Foundation Grant ATM-8609786*, Hazards Assessment Laboratory, Colorado State University, Fort Collins, CO.

Farhar, B.C., and Mewes, J. (1975). "Weather Modification Decision Making: State Law and Public Response." *J. Appl. Meteorology*, **14**, AMS, Boston, MA, 694–701.

Farhar, B.C., and Rinkle, R. (1977). "A Societal Assessment of the Proposed Sierra Snowpack Augmentation Project." Human Ecology Research Services, Inc., Boulder, CO.

Foehner, O.H., Jr., (1983). "Section 2 – Social, Economic, and Environmental Issues in Weather Modification," Part of Guidelines for Cloud Seeding to Augment Precipitation, IR, V. 109, No. 1, ASCE, Reston, VA, March, pp. 116–127.

Garcia, P., Changnon, S.A., Jr., and Pinar, M. (1990). "Economic Effects of Precipitation Enhancement in the Corn Belt." *J. Appl. Meteorology*, **29**, AMS, Boston, MA, 63–75.

Garcia, P., Offutt, S.E., and Pinar, M. (1987). "Corn Yield Behavior: Effects of Technological Advances and Weather Conditions." *J. Climatology Appl. Meteorology*, **26**, AMS, Boston, MA, 1092–1102.

Gelt, J. (1992). "Weather Modification, a Water Resource Strategy to Be Researched, Tested Before Tried." *Arroyo*, **6**, (1), Water Resources Research Center, University of Arizona, Tucson, AZ.

Griffith, D.A., and Solak, M.E. (1999). "A Cloud Seeding Program to Enhance Hydroelectric Power Production from the El Cajon Drainage, Honduras." Proceedings, Seventh WMO Scientific Conference on Weather Modification, Chiang Mai, Thailand, WMP Report No. 31, 131–134.

Griffith, D.A., and Solak, M.E. (2002). "Economic Feasibility Assessment of Winter Cloud Seeding in the Boise River Drainage, Idaho." *J. Weather Modification*, 34, WMA, Fresno, CA, 39–46.

Haas, J.E. (1974). "Sociological Aspects of Weather Modification," in *Weather and Climate Modification*, W.N. Hess, (ed.), John Wiley & Sons, NY, 787–811.

Haas, J.E., Boggs, K.S., and Bonner, E.J. (1972). "Science, Technology, and the Public: The Case of Planned Weather Modification." *In Social Behavior, Natural Resources and the Environment*, W. R. Burch, Jr., N. H. Cheek, Jr., and L. Taylor, (eds.), Harper and Row, NY, 151–173.

Harper, K.T. (ed.) (1981). "Potential Ecological Impacts of Snowpack Augmentation in the Uinta Mountains, Utah." *Final Report to the Water and Power Resources Service*, Brigham Young University, Provo, UT.

Henderson, T.J. (2003). "New Assessment of the Economic Impacts from Six Winter Snowpack Augmentation Projects." *J. Weather Modification*, **35**, WMA, Fresno, CA, 41–44.

Keyes, C.G., Jr. (2002). "Recommendations for Future New Mexico Atmospheric Water Management Projects," Final Report to the New Mexico Interstate Stream Commission, May.

Klein, D.A. (ed.) (1978). *Environmental Impacts of Artificial Ice Nucleating Agents.* Dowden, Hutchinson, and Ross, Inc., Stroudsburg, PA.

Knight, D.H., Anderson, A.D., Baxter, G.T., Diem, D.L., Parker, M., Rechard, P.A., Singleton, P.C., Thilenius, J.F., Ward, A.L., and Weeks, R.W. (1975). "The Medicine Bow Ecology Project." *Final Report to Bureau of Reclamation*, University of Wyoming, Laramie, WY.

Krane, S. (1976). "Social Implications of the National Hail Research Experiment: A Longitudinal Study." *Final Report to the National Center for Atmospheric Research*, Human Ecology Research Services, Boulder, CO.

Lambright, H.W. (1972). "Governmental and Technological Innovation: Weather Modification as a Case in Point." *Public Administration Rev.*, **32**, 1–10.

Larson, W.L. (1973). "Social Impacts, Knowledge, Attitudes, and Expectations of Farmers and Ranchers in Eastern Montana toward Weather Modification and Impacts of Induced Rainfall." *Bulletin* **670**, Montana Agriculture Experiment Station, Montana State University, Bozeman, MT, 25–30.

Lease, J.C. (1985). "Estimated Effects from Cloud Seeding in the Colorado River Basin." *Preprints, Fourth WMO Scientific Conference on Weather Modification*, A Joint WMO/IAMAP Symposium, Honolulu, HI, World Meteorological Organization, Geneva, Switzerland, WMO/TD-No. 53, 463–466.

Loehr, W., Welles, B.C., and Sherretz, L.A. (1983). "The Value of Electric Power and Possible Effects of Weather Modification on Small-Scale Hydroelectric Production in Colorado." *Report, Bureau of Reclamation Cooperative Agreement No. 3-FC-81-01240*, Colorado Department of Natural Resources, Denver, CO.

MacDonald, L.H. (1986). "Persistence of Soil Moisture Changes Resulting from Artificially-Extended Snowmelt." *Proceedings, Western Snow Conference*, Phoenix, AZ, 146–149.

Mewes, J. (1977). "Hail Suppression in South Dakota," in *Hail Suppression: Society and Environment*, Institute of Behavioral Science, University of Colorado, Boulder, CO, 46–60.

Reinking, R.F. (1992). "The NOAA Federal/State Cooperative Program in Atmospheric Modification Research: A New Era in Science Responsive to Regional and National Water Resources Issues." *Preprints, Symposium on Planned and Inadvertent Weather Modification*, Atlanta, GA, AMS, Boston, MA, 136–144.

Reinking, R.F., and Meitin, R.J. (1989). "Recent Progress and Needs in Obtaining Physical Evidence for Weather Modification Potentials and Effects." *J. Weather Modification*, **21**, WMA, Fresno, CA, 85–93.

Reinking, R.F., Berg, N.H., Farhar, B.C., and Foehner, O.H., Jr. (1995). Section 2 of Manual 81, "Economic, Environmental, and Societal Aspects of Precipitation

Enhancement by Cloud Seeding," *Guidelines for Cloud Seeding to Augment Precipitation*, ASCE, Reston, VA, pp. 9–47.

Rogers, E.M. (1983). *Diffusion of Innovations*, Free Press, NY.

Rogers, E.M., and Shoemaker, F.F. (1971). *Communication of Innovations*, Free Press, NY.

Romm, J.F., and Ewing, A. (1987) "The Economic Value of Water in National Forest Management." *Proceedings, California Watershed Management Conference*, **11**, R. Z. Callahan and J. J. Devries, (eds.) Wildland Resources Center, University of California, Berkeley, CA 89–102.

Schaffner, L.W., Johnson, J.W., Vruegdenhl, H.F., and Enz, J.W. (1983). "Economic Effects of Added Growing Season Rainfall on North Dakota Agriculture." *Agricultural Economics*, **172**, Department of Agricultural Economics, Agricultural Experiment Station, North Dakota State University, Fargo, ND.

Sewell, W.R.D. (ed.) (1966). "Human Dimensions of Weather Modification," **105**, Department of Geography Research, University of Chicago Press, Chicago, IL.

Sewell, W.R.D. (1969). "Weather Modification: When Should We Do It and How Far Should We Go?" In *Weather Modification, Science and Public Policy*, R.G. Fleagle, (ed.), University of Washington Press, Seattle, WA.

Sherretz, L.A. (1983). "Early-Season Snow and Skier Visits in Colorado." *Weather Modification Regulation and Research*, Colorado Department of Natural Resources, Denver, CO.

Sherretz, L.A. (1984). "Daily Snowfall and Skier Visits in Colorado." *Report, Bureau of Reclamation Cooperative Agreement No. 4-FC-81-04620*, Colorado Department of Natural Resources, Denver, CO.

Sherretz, L.A., and Loehr, W. (1983). "A Simulation of the Costs of Removing Snow from County Highways in Colorado." *Report, Bureau of Reclamation Cooperative Agreement No. 1-07-1981-V0226*, Colorado Department of Natural Resources, Denver, CO.

Smith, F.W. (1978). "A Study of the Effects of Altering the Precipitation Pattern on the Economy and Environment of Kansas." *Final Report to the Kansas Water Resources Board*, Kansas State University Agricultural Experiment Station, Manhattan, KS.

Smith, J.L., and Berg, N.H. (1979). "Biography of the Environmental Effects of Weather Modification." *The Sierra Ecology Project*, **4**, Office of Atmospheric Resources Management, Bureau of Reclamation, USDI, Denver, CO.

Smith, J.L., Canine, N., Dix, R.L., Huntington, G.L., Major, J., Patrick-Reid, C., Scherrell, A., and Steinhoff, H. (1978a). "An Evaluation of Possible Effects of Weather Modification upon Vegetation in the American River Basin, California." *The Sierra Ecology Project*, **1**, (2), Office of Atmospheric Water Resources Management, Bureau of Reclamation, USDI, Denver, CO.

Smith, J.L., Longhurst, W.M., Neal, D.L., Salwasser, H.J., Schultz, A.M., and Zinke, P.J. (1978b). "An Evaluation of Possible Effects of Weather Modification upon Deer and Their Habitat in the American River Basin, California." *The Sierra Ecology Project*, **1** (3), Office of Atmospheric Water Resources Management, Bureau of Reclamation, USDI, Denver, CO.

Smith, J.L., Ferrell, G.T., Koerber, T.W., Parmeter, J.R., Scharpf, R.F., and Srago, M.D. (1978c). "An Evaluation of Possible Effects of Weather Modification upon Forest Insects and Diseases in the American River Basin, California." *The Sierra*

Ecology Project, **1** (1), Office of Atmospheric Water Resources Management, Bureau of Reclamation, USDI, Denver, CO.

Smith, J.L., Erman, D.C., Hart, D.D., Kelly, D.W., Klein, D.A., Koch, D.L., Linn, J.D., Moyle, P.M., Ryan, J.H., and Woodard, R.P. (1980). "An Evaluation of Possible Effects of Weather Modification on Lake and Stream Biota in the American River Basin, California." *The Sierra Ecology Project*, **2**, (5), Office of Atmospheric Water Resources Management, Bureau of Reclamation, USDI, Denver, CO.

Sonka, S.T. (1979). "Economics of Weather Modification: A Review." *Report of Investigation*, Illinois State Water Survey, Urbana, IL.

Stauffer, N.E., Jr. (2001). "Cloud Seeding—The Utah Experience." *J. Weather Modification*, **33**, WMA, Fresno, CA, p. 68.

Stauffer, N.E., and Williams, K. (2000). Utah Cloud Seeding Program Increased Runoff/Cost Analyses. Utah Department of Natural Resources, Division of Water Resources, Salt Lake City, Utah.

Steinhoff, H.W., and Ives, J.D. (1976). "Ecological Impacts of Snowpack Augmentation in the San Juan Mountains." *Final Report to the Bureau of Reclamation*, Colorado State University, Fort Collins, CO.

Stroup, R.L. (1973). "Economic Impact (Weather Modification)." *Bulletin* **670**, Montana Agriculture Experiment Station, Montana State University, Bozeman, MT, 31–36.

Super, A.B., and Reynolds, D.W. (1991). "The Feasibility of Enhancing Streamflow in the Sevier River Basin of Utah by Seeding Winter Mountain Clouds." *Report No. R-91-10*, Bureau of Reclamation, USDI, Denver Federal Center, Denver, CO.

USDA. Forest Service and State of California Department of Water Resources. (1990). "Joint Environmental Impact Statement Environmental Impact Report—Prototype Project to Augment Snowpack in Plumas and Sierra Counties." Plumas National Forest, Forest Service, USDA, Quincy, CA.

USDA (1991). "Final Supplement to the Environmental Impact Statement for the Prototype Project to Augment Snow Pack by Cloud Seeding Using Ground-Based Dispensers in Plumas and Sierra Counties." Plumas National Forest, Forest Service, USDA, Quincy, CA.

USDI (1977a). "Final Environmental Impact Statement, Project Skywater." Report No. *Int Fes 77–39*, **1**, (9), Bureau of Reclamation, USDI, Denver Federal Center, Denver, CO.

USDI (1977b). "An Overview of the Skywater IX Conference on Precipitation Management and the Environment: Discussion and Summary Reports." Engineering Research Center, Bureau of Reclamation, USDI, Denver, CO.

Walkinshaw, H. (1985). "The Economic and Social Impacts of Long-Term Federal and State Supported Weather Modification Projects in the United States." *Report to the North Dakota Atmospheric Resource Board*.

Weisbecker, L.W. (1974). "The Impacts of Snow Enhancement–Technology Assessment of Winter Snowpack Augmentation in the Upper Colorado River." *Stanford Research Institute*, Menlo Park, CA, University of Oklahoma Press, Norman, OK.

SECTION 3

LEGAL ASPECTS OF WEATHER MODIFICATION OPERATIONS

George W. Bomar[7]*, A.M. ASCE*

3.1 INTRODUCTION

For a society to reflect the will of the people and accommodate their competing interests, laws governing their behavior must be expressions of policy decisions intended to ensure an equitable balance between the opportunity to advance individual, and collective, interests and the need to consider the rights of others. Law is a tool through which such determinations about public policy are expressed and applied (Hurst, 1950). This chapter has been adapted from Davis (1995) with adjustments to reflect recent developments pertaining to legal implications of the use of cloud seeding technology.

Laws pertaining to weather modification, as with other activities, are formulated and implemented by all three branches of government. The U.S. Congress enacted Public Law 92-205 [Title 15 U. S. Code Anno. section 330-330(e) (West 1976 & Supp. 1993)] to require individuals and organizations performing weather modification activities to report them systematically so the public's right to know is accommodated. Various states, to protect citizens from incompetent, or dishonest, purveyors of weather modification technology, enforce an array of weather modification rules and regulations. [See Section 3.7, Appendix of (Davis, 1995) for a list of state statutory and regulatory references.]

Such statutes most often merely express policy in the broadest sense. The responsibility to "fill in the details" rests with administrative entities, such as state agencies, which possess rule-making authority. For instance, the National Oceanic & Atmospheric Administration is the federal agency with the power to make policy about reporting on weather modification activity. Rules and regulations that have to do with professional licensing

[7]Senior Meteorologist, Texas Department of Licensing and Regulation, Austin, Texas 78711.

and permitting within individual states are promulgated by pertinent state regulatory agencies. [Refer to Section 3.7, Appendix of Davis (1995) for additional information.]

While their law-making role is not nearly as overt, the courts have essentially formulated law relating to atmospheric water rights and cloud seeding liabilities. This has been accomplished through case decisions made by the courts, including the writing of opinions justifying such decisions, thereby establishing precedents which often are followed in subsequent, similar court cases. Thus, any understanding of weather modification law must be gained from both judicial actions as well as the work of the other two branches of government.

The ability of the federal government to impact weather modification policy resides within various appropriation acts which make government funding available for research in, and development of, cloud seeding technologies. Such influence was first brought to bear in the 1970s, when Congress not only enacted a law calling for a study of prospects and policies for atmospheric water management but also furnished the funding for such a study. The investigation was duly carried out and reported (Weather Modification Advisory Board, 1978). Nonetheless, the only federal statute governing weather modification has to do with reporting activity, and practically all of American law specifically targeting weather control activities rests with the states.

The legal aspects of atmospheric water-resource development and implementation are at play during three different phases in the life of a weather modification project: (1) pre-operational planning; (2) the conduct of a project; and (3) the evaluation of the work accomplished.

3.2 PRE-OPERATIONAL PLANNING

The steps to be taken in preparing for a weather modification project usually include, but may not be limited to: (1) obtaining a weather modifier's professional license; (2) securing a weather modification operational permit; (3) complying with environmental law stipulations, where relevant; and (4) entering into a contract of sponsorship between the entity sponsoring the weather modification operation and the individual, or organization, conducting the project.

3.2.1 Role of Regulatory Entities in the States

Nearly two-thirds of the states have, at some time and to varying degrees, implemented legislation about cloud seeding. The scope of these state laws varies from the rather complete North Dakota regulatory and funding scheme to the bare mention of atmospheric water in Hawaii (Beck, 1991). In the vast majority of the states with weather modification regulatory responsibility, provisions exist to require people intent on doing weather modification work to submit to a regimen for gaining the necessary credentials.

3.2.2 Weather Modification Licenses

Since occupational competency is of fundamental public concern, the licensing of Professionals in the United States is commonplace. Like doctors, lawyers, accountants, and engineers, those who practice weather modification must be licensed (including the regulators).

3.2.2.1 A Basis for Occupational Competency. Licensing criteria focus on two major factors: educational qualifications and operational experience. Generally, these are merely alluded to, if not mentioned, in various states' legislation. It is up to agencies implementing the statutes to ensure that they are fleshed out in administrative rules. Although competency criteria differ among the states, they most often list a minimum number of academic credit hours of college instruction in meteorology, mathematics, engineering, and other physical sciences, or combinations of these. They usually also require a certain number of years of experience working at weather modification projects. As a rule, integrity requirements are not so specifically delineated. However, at the very least, the criteria, when met, suggest the licensee has not been shown to be dishonest. A decision to license a meteorologist for work in a particular state sometimes relates to whether that individual has been certified by the Weather Modification Association (Weather Modification Association, 2004a).

3.2.2.2 Initiating the Process for Obtaining a License. Usually, where applications for weather modification licenses are concerned, licensing agencies have discretionary authority in deciding whether to grant, or deny, a license. Administrators have more than the power to determine whether the application process has been fully complied with. They can weigh qualifications as part of their decision-making process. Unless administrative fact-finding is arbitrary or unreasonable, regulators have the final word on who can practice weather modification in their state (Davis, 1995). Courts will not overturn licensing decisions just because the judges disagree with the regulators (Pierce et al., 1985).

While some state regulatory agencies have the prerogative to call for an interview of the applicant, most often face-to-face meetings are not necessary. Instead, regulators use information submitted by the applicant in the form of standardized applications, accompanying documents (e.g., diplomas, college transcripts, licenses from other states, lists of published papers, descriptions of work experience), as well as letters of recommendation, and data otherwise unavailable to the licensing agency. Testing, like that employed by legal or accountancy board certification examinations, is rarely done.

Weather modification licenses most often are good for one year, and license fees in most states are quite modest–a few hundred dollars at most per year. Renewal in most instances is virtually automatic.

Suspension, or revocation, of licenses, as well as refusal to renew, is all theoretically possible under most state weather modification statutes and agency rules. As a practical matter, though, the initial grant of the license (or its denial) is the most critical decision made by the regulator.

3.2.3 Weather Modification Permits

Governmental regulation of weather modification not only ensures that the interests of the public are protected by a somewhat vigorous licensing effort, but it also supports the implementation of the technology for the public's benefits. Permitting is designed to protect the public interest by excluding projects with inadequate technical merit and insufficient financial backing. By weeding out the unsavory projects, the worthwhile ones are helped. Moreover, permits are designed to ensure that adjoining projects do not interfere with one another.

Nearly half of all the states have in place regulatory provisions that require government approval of the "conceptual plan" of a weather modification project before cloud seeding can be lawfully carried out (Table 3-1).

In most cases, state weather modification laws provide for exemptions to certain activities. These could include the following situations.

3.2.3.1 Criteria Considered during the Permit Process. Before issuing a weather modification permit, administrative agencies in various states consider the following and want the answers to the questions listed:

1. *Personnel:* Will the project have licensed personnel on the premises? Are the meteorologist(s), pilot(s), and other employees well qualified for their respective roles?
2. *Methodology:* What cloud-seeding agents will be used? What are the rate, timing, and method(s) of dispersal?
3. *Equipment:* What types of dispensing gear will be employed (ground-based generators, aircraft)? Are the weather radar and other monitoring and measuring equipment adequate for the tasks planned?
4. *Project area:* Are the target, operational, and control areas well delineated?

TABLE 3-1. States with Cloud Seeding Regulations, Legal Rights and Liabilities Provisions (Davis, 1995; Keyes, 2003; New Mexico Law, 2003; Langerud, 2004; Solak 2004)

States	Public Funding	Project Regulation	Notice of Project	Reporting Seeding	Legal Rights and Liabilities
Arizona		X		X	
California	X		X	X	X
Colorado	X	X	X	X	X
Florida		X		X	
Idaho	X		X	X	
Illinois	X				
Iowa	X				
Kansas	X	X		X	
Louisiana	X	X			
Michigan		X		X	
Minnesota	X	X		X	
Montana	X	X		X	
Nebraska	X	X			
Nevada	X	X		X	
New Hampshire	X				
New Mexico	X	X	X	X	
New York	X				X
North Dakota	X	X	X	X	X
Oklahoma	X	X	X	X	
Oregon	X	X		X	
Pennsylvania		X		X	X
South Dakota	X	X		X	
Texas	X	X	X	X	X
Utah	X	X	X	X	X
Washington	X	X		X	
Wisconsin		X		X	
West Virginia		X		X	X
Wyoming	X	X	X	X	

Explanation: Public Funding—government funding of cloud seeding projects
Project Regulation—requiring project permit and/or cloud seeding license as a condition of lawful seeding
Notice of Project—requiring publication of a notice of the project
Reporting Seeding—record keeping of projects and periodic reports of them
Legal Rights and Liabilities—legal liability for harm caused and legal rights in water

5. *Operations plan:* Are the methodologies to be used of sound, scientific merit? Are there cloud-seeding suspension criteria to be followed?
6. *Data collection and archival:* What are the plans to obtain, process, and analyze data to evaluate the efficacy of the project?
7. *Assessment:* What methods will be used to evaluate the impact of cloud seeding? Will the planned work impact any other permitted weather modification operation?
8. *Contract and cost information:* What contracts will be negotiated?
9. *Liability insurance:* What types of insurance coverage will be furnished? Is such insurance adequate?

By issuing permits either as requested or in altered form, regulators are able to shape projects in ways best suited to protect the interests of the cloud seeder, the sponsor, people affected by the project, and the public. Many states have highly competent agency personnel who carry out their delegated tasks in a highly professional manner. These personnel may rely on advisory committees, or boards of experts, for needed expertise in processing requests for permits. In states where effective regulatory work is in doubt, the deficiency may well be due to inadequate funding.

Once an application for a permit is in hand, the relevant state agency decides what actions to take. Its strategy may be aided by the fact that its personnel already possess familiarity with the design of the project to be permitted. In most states, one requirement of the permit applicant is that public notice of the permit application be published in newspapers in the area to be impacted by the project. This enables the public to bring any concerns about the proposed project to the attention of the agency. The agency may determine that public hearings should be conducted to facilitate interaction with the public on various permit issues.

3.2.3.2 Site and Time Specificity. In most instances a weather modification permit is issued for a single project for a specified time period (from one season to several years in duration). Consequently, any appreciable delays in making permitting decisions can adversely impact projects. Decisions made by state agencies are subject to judicial review, and courts usually uphold administrative determinations. However, in cases involving errors of law or unreasonable findings of fact, courts may reverse agency decisions (Gellhorn and Levin, 1990). Since long-term weather modification projects often involve substantial financial commitments, in some states regulatory agencies are given power to grant provisional approval (subject to annual review) to permit applications for longer periods of time. Permit fees are either set by state law as fixed amounts or as percentages of contracts.

The initial decision by an agency to grant, or deny, a permit is almost invariably the critical one. However, there are instances in which regulators refused to renew permits, and there is authority in regulating agencies in some states to modify permits while they are in force. A permit is a valuable right but one which can be administratively altered, or expired, at the end of its designated term.

3.2.3.3 Historical Overview of Permitting Controversies. In recent years a number of permitting disputes have been particularly vigorous. Local opposition to hail suppression in the high plains of Texas led either to denials or revocation of weather modification permits, which in turn resulted in alterations to the Texas regulatory law making it easier for protesters to block permit issuance where hail suppression is the objective (Kirby, 1978). An appeal to the courts for judicial review of a permit decision in Colorado resulted in a decision by the regulating agency to not consider permit applications unless, and until, it was given a line-item appropriation to enforce its cloud seeding control law (Davis, 1987).

Litigation headed off issuance of a permit in Montana for snowpack augmentation to the Bonneville Power Administration [*Montana Wilderness Association v. Hodel*, 380 F. Supp. 879 (D. MT 1974)]. Then, in 1992, litigation over the effort by operators of the North Dakota weather modification program to seed clouds resulted in issuance of a permit, ordered by a Montana court, to allow seeding inside Montana upwind from its North Dakota target area [North Dakota Atmospheric Resources Board v. Board of Natural Resource and Conservation, No. ADV-92-918 (MT 1st Judicial District Court, 1992)] (Davis and St. Amand, 1975).

3.2.4 Impacts of Environmental Laws and Rules

With its goal of altering natural weather intentionally, the management of atmospheric water attracts scrutiny from people concerned over environment quality. Sponsors and operators of weather modification projects must comply with applicable environmental statutes and regulations.

3.2.4.1 Adherence to Environmental Constraints. State laws requiring a valid permit as a condition of lawful cloud seeding can be used to further environmental protection goals. During the permitting process, regulators can ascertain whether a project is environmentally sound (Davis, 1995). Then, as cloud seeding projects are conducted, the reporting of activities (usually by the agency or some contractor working for the regulator) yields information that can be used to determine whether the projects have complied with environmental considerations (Davis, 1975).

Where compliance with environmental laws is in question, public or court action can affect remedies. Possible changes in snowpack depth, for example, triggered expression of concern over wildlife by a Montana environmental group that went to court to stop the proposed cloud seeding project (Davis, 1995). The group invoked environmental laws that it asserted would have been violated [*Montana Wilderness Association v. Hodel, 1974* (Davis, 1968)].

3.2.4.2 Considerations for Environmental Impact Statement. Federal agencies proposing to undertake projects having a "significant" impact upon the "quality of the human environment" are required by the National Environmental Policy Act of 1969 to file a "detailed" Environmental Impact Statement (EIS) [Title 42 U.S. Code Anno. section 4321 et. seq., Pub. L. 91-190 (1970)]. A few states, including some with active weather modification projects, have passed similar laws applying to state agencies. Because funding research and development of weather modification technology has been an important federal activity and using government money for operations has been undertaken, in some jurisdictions, by the state, its subdivisions, or both, requirements for impact statements can be important (Davis, 1995). Such statements require appreciable time and money, and if they are not done correctly, the project could be enjoined until the EIS is acceptable.

Laws mandating environmental impact statements require that the EIS contain: (1) the environmental impact of the proposed action; (2) any adverse environmental effects which cannot be avoided should the proposal be implemented; (3) alternatives to the proposed action; (4) the relationship between local short-term uses of man's environment and the maintenance and enhancement of long-term productivity; and (5) any irreversible and irretrievable commitments of resources that would be involved if the proposed action should be implemented (Plater et al., 1992; Davis, 1995).

Preparing an EIS usually follows certain steps: (1) data collection concerning the elements of the EIS, which may be by "borrowing" from other statements relating to similar projects, by carrying out an "environmental assessment," or by conducting studies; (2) preparation of a draft impact statement; (3) circulation of the draft statement for receipt of comments from interested governmental agencies, groups, and persons; (4) consideration of comments, and reaction to them, by altering the proposal or the final EIS, or both; and (5) filing the final statement with the appropriate governmental agency (Anderson, 1973; Davis, 1995).

There are cases in which snowpack augmentation has run afoul of bureaucratic interpretation of federal environmental policy. The Wilderness Act of 1964 established the Wilderness System, units of which are added by specific congressional inclusions of parcels of federal lands deemed appropriate for wilderness protection (Davis, 1995). In wilderness areas, where many prime sites for snowpack augmentation exist, certain activities are banned. Officials of some federal agencies have taken the position that the law bars installation and monitoring of hydrometeorological data-collection equipment, at least where mechanized means are used to install and service them (Davis, 1975 and 1995).

3.2.5 Contractual Agreements among Sponsors and Operators

Where a sponsor seeks to enlist the services of an individual or organization to conduct cloud seeding operations, a key step in the process is negotiation, preparation, and execution of a formal sponsorship contract.

3.2.5.1 Perspectives of Sponsors and Operators. From the operator's perspective, a critical concern about the prospective sponsor is ability to pay the contract price. Some sponsors, such as utilities and ski areas, have adequate financial resources. So do governmental sponsors, though they must rely on the appropriations process of legislatures and comply with relevant fiscal laws in order to raise and spend taxpayers' money. On the other hand, entities without ad valorem taxing authority, like agricultural cooperatives, have a less attractive record of being able to fund projects over an extended period of time.

Reliance on governmental funding of weather modification has its drawbacks. Where taxpayers' funds for public works projects like cloud seeding are concerned, three types of funding laws applicable to atmospheric water management are: (1) general legislative grants of authority to spend appropriated funds; (2) authority specially granted to the sponsoring agency by law to spend money on cloud seeding activities; and (3) such special authority coupled with power to establish legal entities that have taxing power to raise funds for weather modification (Changnon et al., 1977). Obviously, the availability of funds for a project is dependent upon political support. In the absence of adequate appropriated monies, the project has to be terminated (Donnan et al., 1976).

From the sponsor's point of view, the ease—or difficulty—with which contracts are successfully negotiated is related to the availability of operators. Historically, the number of eligible individuals, or organizations, with the requisite experience and tools to do cloud seeding is small. Personnel whose services are contracted for should have a weather modification license, if the project is located in a jurisdiction requiring licensing. The Weather Modification Association maintains information about certified operators and project managers, who are listed in its publication (page 107), the *Journal of Weather Modification* (Weather Modification Association, 2004b) and on the web site, www.weathermodification.org, maintained by the Association.

Contracts negotiated between sponsors and operators address the following concerns: Who will be responsible for obtaining the necessary licenses and permits? Who will supply needed weather forecasts and other relevant meteorological services? What seeding material delivery systems will be furnished, and when? What records, including cloud-seeding data, weather information, and hydrological data, will be maintained and archived? What suspension criteria will be followed, and how will sponsors and operators convey their concerns about implementation of those criteria? Who will be responsible for writing, and submitting to the proper authorities, the requisite summary reports? Who will arrange for legal liability insurance coverage?

3.2.5.2 Ethical Standards Relating to Expectations and Claims. As recommended by the ethics and standards policies of the Weather Modification Association, operators should take measures to comply with professional ethical standards relating to sponsorship contracts (Weather Modification Association, 2004c). Operators should not exaggerate their capabilities nor guarantee results. Nor should they ever contract for bonuses based on producing precipitation over and above certain thresholds, such as monthly normal or other arbitrary amounts. Such assertions and arrangements are detrimental to the weather modification industry. To say it another way, contingent fee contracts are never appropriate.

3.3 CONDUCTING OPERATIONS

Providers of cloud seeding services have certain legal obligations to meet during those times when weather modification activities are underway. These include: (1) operating within the parameters, and constraints, of the approved operational plan(s); (2) maintaining accurate and thorough records of their activities and the results; and, (3) reporting in a timely way about their activities to sponsors, the National Oceanic & Atmospheric Administration, and relevant state regulatory agencies.

3.3.1 Operational Control

Great care should be exercised by those performing cloud seeding services. One reason for diligence is obvious: Every operation should be conducted in ways that avoid harming people or property, and that facilitate the successful pursuit of the objective (e.g., precipitation augmentation). But another reason has to do with avoiding the cultivation of any public perception that harm has, or will, result from cloud seeding or that no benefit has been, or can be, derived from such activity.

Those who manage weather modification projects can aid themselves, and sponsors, in avoiding these pitfalls by using public advisory groups and by operating only under predetermined (sometimes strict) operational controls. Such controls are often stipulated in the weather modification contract between sponsor and operator as well as in relevant permits. The controls consist of a clear designation of target and operational areas, a listing of cloud seeding criteria being observed, and a description of the policies and procedures in place to ensure that suspension criteria are observed when conditions warrant. Such controls should not, however, be so restrictive that they inhibit, or prevent, the successful capitalization of safe cloud seeding opportunities (Bluestein et al., 1986). The kinds of record keeping and reports required by contracts between cloud seeders and sponsoring groups will usually reflect the degree to which the project has conformed to the operational criteria established by the parties and regulators.

3.3.2 Archival of Data and Information

Anyone paying for cloud seeding services has a right to know what seeding materials and methodologies have been used and what results are believed to have been realized as a consequence of the weather modification activity (Weather Modification Association, 2004c). After all, information is the lifeblood of effective government regulation, as well as the key to keen public awareness of what is happening within a regulated industry like weather modification.

Accordingly, the operator must keep accurate and exhaustive records from which he/she can demonstrate compellingly that contractual obligations have been met. Such records will almost always satisfy regulators and address the dictates of government red tape and paperwork.

Officials representing the regulatory agencies have authority to visit on site to obtain information about the project. In some instances, government employees may linger on site to monitor day-to-day operations, though more often visits are made on an intermittent, even hit-or-miss, basis. The visitorial power given to regulators by state laws is only used as an adjunct to police the record keeping and reporting requirements (Gellhorn and Levin, 1990). The type of reporting routinely done by the weather modifier, which entails gathering the data and entering it on proper forms, is less expensive and also less intrusive.

3.3.3 Reporting Procedures

Federal reporting is intended merely as a disclosure requirement. The periodic reports from cloud seeding operators to the National Oceanic & Atmospheric Administration are compiled and then disseminated by the federal agency in a comprehensive annual report on all weather modification activity in the United States. The report is then made available to the public. The federal government takes no action on the reports since it is not in the regulatory business.

Federal reporting is for information purposes only, not as evidence of any evaluation done on the project. Cloud seeding activities are reported, not the impacts from the seeding. Following submittal of an initial report to establish a federal record of a seeding project, the regulations, Part 908 of NOAA rules, require both interim (seasonal/annual) and final reports upon completion of projects. These reports to NOAA include: (1) the number of days each month when seeding operations were conducted, and for what purposes (rain or snow enhancement, hail suppression, fog dispersal); (2) hours of operation of each apparatus (airborne or ground-based) used in the project; and (3) the types, and amounts, of cloud seeding agent used [Title 15 Code Federal Regulations section 908.8 (1987 & Supp. 1991)].

In some states, the federal reporting forms suffice as the documentation required by state regulators. Other states insist on their own reporting forms, which usually are quite similar to the federal forms. The information provided by operators is used by regulators to verify conformance to provisions of the relevant licenses and permits, as well as to keep the public fully informed. Those states without cloud seeding regulations do not require any reporting of activity.

3.4 EVALUATING OPERATIONS

Most law dealing with legal liabilities of weather modification activities to people claiming harm from weather modification activities is judge-made law, which is developed by courts following precedent set down in analogous situations. There is a large body of such so-called common law relating to liability for harm. Most of it comes from state courts, or from federal courts applying state law (Davis, 1995). Although the number of weather modification lawsuits historically has been few, by drawing on liability law developed from similar situations, a picture can be deduced of potential liabilities for damages for cloud seeding activities (Davis, 1974).

3.4.1 Legal Liabilities for Sponsors and Operators

At present, only eight states have law dealing specifically with weather modification legal rights and liabilities. In five jurisdictions, there has been at least one case in court involving issues about atmospheric water rights or weather modifiers' liabilities (see Table 3-2). Although weather modifiers have been successful in responding to liability challenges, well-advised cloud seeders carry legal liability insurance. Such liability coverage may be required by statutes and contracts (Davis, 1995). Legal defense expenses as well as judgments are payable from such insurance policies (Dobbyn, 1989). Such costs can be quite substantial (Mann, 1968).

Statutes on legal liability change the common law. In two states, statutes limit the theories upon which liability may be based by limiting plaintiffs to suing for professional malpractice or intentional wrongdoing. This is done by stating that the cloud seeding is not an ultra-hazardous activity for which there is liability without fault, or by stipulating that there can be no liability either for trespass or nuisance merely for inserting cloud seeding agents into the atmosphere. Pennsylvania and West Virginia go the other way. Statutes in those states make proving liability cases easier by dropping the requirement that claimants establish that cloud seeders were at fault. Plaintiffs merely need to prove that they were harmed by the activity of the cloud seeder (Davis, 1995) (see Table 3-2).

3.4.1.1 Liability Theories. It is not unusual for plaintiffs' lawyers to allege all of the liability theories. Proof of such allegations, however, is a real challenge. Plaintiffs have the burden of bringing evidence to court that will persuade the finder of fact that the defendants' conduct met the requirement of at least one such theory. To date, defendants have won almost all liability lawsuits.

One example of the difficulty faced by plaintiffs' lawyers in prevailing in liability cases is the *Yuba City Flood Case,* instituted in California during the 1950s and concluded the following decade (Davis, 1995). Complainants alleged several theories on the basis of which they sought to have the court determine that the cloud seeding in question was the sort of inappropriate action on which the law would rest liability [*Adams v. California,* Civil No. 10112 (Superior Court Sutter County, CA, 1964)]. They asserted that the cloud seeders: (1) were negligent (that their conduct was professional malpractice because it fell below the standard of conduct expected of professional cloud seeders); (2) trespassed (that they caused intrusion

TABLE 3-2. State Cloud Seeding Rights and Liabilities Provisions (Davis, 1995)

	Provisions
	Case Law on Atmospheric Water Rights and Liabilities
New York	Cloud seeding proper because property owner has no atmospheric water rights.
Texas	Operator can be liable because property owner has atmospheric water rights.
Pennsylvania	Cloud seeding proper if government-authorized.
	Statute on Rights in Augmented Water
Colorado	Water can be appropriated.
Utah	Water use right is with next person with an unfilled water right.
North Dakota	Treatment as if it were natural water.
	Statute on Liability Theory
Texas	Cloud seeding is not regarded as an ultra-hazardous activity.
Utah	Cloud seeding liability possible only for negligence.
Pennsylvania	Liability if defendant's conduct harms plaintiff (no fault).
West Virginia	Liability if defendant's conduct harms plaintiff (no fault).

of materials, rain/snow, and runoff, or combinations of these, on lands owned by the plaintiffs; (3) committed a private nuisance (that on the balance, the gravity of harm from their conduct outweighed the benefit to the cloud seeders and their sponsor); and/or (4) performed an abnormally dangerous activity (that cloud seeding is so dangerous that its performance should be liable for harm caused by them, even though they may not have been at fault by being careless, trespassing, or committing a nuisance) (Mann, 1968). Proof of at least one of these liability theories is one of the necessary elements of a plaintiff's liability case (Keeton et al., 1984).

3.4.1.2 Causation. It is not enough for people seeking to recover damages in court for misconduct by others to allege, and prove, some theory of liability. Rather, it must also be shown that some causal connection exists between the conduct on which that theory is based and the harm which they assert has befallen them. It is on this requirement of proof of causation that most weather modification plaintiffs have floundered. For example, in a Michigan lawsuit, a farmer claimed that cloud seeding in an intended target area upwind from his farm caused a storm which had an adverse effect on his property [*Reinbold v. Sumner Farmers, Inc.*, No. 2734C (Tuscola County, MI, 1974)]. The jury, which found for the defendants, evidently concluded that either he failed to establish a liability theory or he had not proved causation (Davis and St. Amand, 1975). Examination of the evidence indicates that the plaintiff did not show causation (Davis, 1995). No wonder, then, why so few lawsuits have been filed. Failure to show causation is a major barrier to any effort to assign legal liability to cloud seeders.

3.4.1.3 Defenses. Those sued for alleged harm from weather modification activities may prevail not only because of the great difficulty of plaintiffs to establish cases. A successful defense may also stem from the defendants proving an affirmative legal defense. The federal government, and a few state governments, is legally immune from liability, at least to the extent they have not waived that immunity. The Federal Tort Claims Act is a partial waiver of federal immunity, but it does not waive liability for abnormally dangerous activities where there has been no governmental fault and for so-called discretionary functions (e.g., project planning) (Jayson, 1984; Keeton et al., 1984). Immunity is an important defense (Davis, 1995).

The Federal Tort Claims Act was the basis on which the plaintiffs in *Lunsford v. United States* [570 F.2d 221 (1977)] sought to obtain a recovery from the federal government for the loss of life and property associated with the Rapid City flood in the early 1970s. Although there was reason to decide the case on its merits against the plaintiffs (Davis, 1988), it actually floundered on the inability of the litigants to have their claims certified as a class action, whereby all of them could join together in a single suit. The procedural provisions of the Federal Tort Claims Act do not permit such a joiner. Those suing under that law must comply with its procedural requirements as well as prove their tort claims. Certification as a class action also has been denied by a state court [*Saba v. Counties*, 307 N.W. 2d 590 (N.D., 1981)].

State defenses include the concept of public necessity, which deals with allowing conduct that might otherwise be the basis for liability if it is necessary to protect the public from an imminent public disaster. Typical cases involve blowing up houses to prevent the spread of conflagration (Kionka, 1992). Might not drought relief by way of cloud seeding also fit?

3.4.1.4 Indemnity and Insurance. The ultimate financial burden of liability can be arranged by contract between the weather modification operator and sponsor or among them and an insurance carrier. Some sponsors (for example, the federal government) may require weather modification operators working for them to agree to indemnify them for any losses incurred, including any legal liabilities. People in the weather modification business can buy legal liability insurance, which will shift the ultimate loss to the insurance carrier (Davis, 1995). State laws and regulations often require liability insurance as a condition of getting a permit. Obviously, the cost of the insurance premiums, like any other costs, gets passed on to the sponsors of the weather modification project.

3.4.2 Water Rights

Practically speaking, liability has seldom been a serious problem for practitioners of weather modification. But the threat of it, and the costs of litigation, continue to be a concern among some operators. The fact is, any time law is broached, those who do cloud seeding think of liabilities. On the other hand, mention cloud seeding to a lawyer and he will think of water rights problems. Yet, water rights problems are often more theoretical than practical. The same reasoning is applicable to water rights questions as that explaining the lack of success by liability claimants, i.e., proving causation. Until better means are available to establish the extent to which clouds have been "rustled" by being seeded and the amount by which cloud seeding efforts have augmented streamflow, it will be quite difficult to quantify whatever right a claimant might be asserting (Davis, 1995). Of course, that day could come, and engineers and other scientists could play active roles in supplying the needed proof and in drafting laws concerning quantification (ASCE, 2004). Nevertheless, there now is some law on the subject of atmospheric water rights (see Table 3-2).

3.4.2.1 Atmospheric Water. Three states (Pennsylvania, New York, and Texas) have case law dealing with ownership of atmospheric waters [*Pennsylvania Natural Weather Association v. Blue Ridge Weather Modification Association,* 44 Penna. District & County Rep. 2d 749 (Common Pleas, Fulton County PA, 1968); *Slutsky v. City of New York, 197* Misc. *730,* 97 N.Y.S.2d 238 (Supreme Court, 1950); *Southwest Weather Research, Inc. v. Duncan,* 319 S.W.2d 940 (TX Civil Appeals Court, 1958)]. Thus, the cases are scattered; they do not come from the top appellate courts of any of the three states, and they are contradictory (Davis, 1995). One case says that the landowner has a right in atmospheric water passing above the surface of his land (Texas); another takes the position that he/she does not (New York); and the third says he/she does but that state-permitted cloud seeding can deprive him of such a right (Pennsylvania) (Davis, 1974). Consequently, it is difficult to deduce any general rule from the cases directly in point.

It is reasonable to analogize rights in atmospheric waters to rights in surface water under the traditional riparian rights system of the eastern states, which would favor property owners holding riparian lands (people with property under the clouds) or under the proper appropriation system of the western states, which support claims by people first making beneficial use of water (sponsors of cloud seeding projects). It is also conceivable that atmospheric water rights might be based on the concepts of developed water, or imported water, which provide the basis for claims by those who bring new water to a basin (Getches, 1990; Tarlock, 1988). Obviously, reasoning from these and other analogies also leads to conflicting results.

3.4.2.2 Augmented Surface Water. The scarcity of atmospheric water rights cases can be attributed to the real concern that water and weather resources management relates to water on the ground. In this instance, engineering measurement and evaluating data present difficult problems. Nonetheless, three states have statutes dealing with such water on the surface of the ground: Colorado (Colorado Legislative Council, 1971), Utah (Dewsnup and Jensen, 1977), and North Dakota. Under Colorado law, a permit can be obtained to appropriate the right to use surface water made available through cloud seeding (Davis, 1995). With Utah law, the right to surface water is allocated to the most senior water appropriator whose allocation was not already filled by water naturally in the stream. North Dakota would treat the additional water like natural runoff (Jones, 1991). Of course, three different solutions by the three states that have addressed the questions are inadequate bases for declaring any trend. A published Regulated Riparian Model Water Code could serve as one basis for a sponsor to claim a quantifiable amount of atmospheric water (ASCE, 2004).

3.5 CONCLUSIONS

Those who develop, and manage, water resources possess considerable experience in dealing with governmental institutions, which both support their activities and administer legal constraints within which waters are administered. Thus, the development of atmospheric water resources tends to be treated consistently with surface and underground water development. Legal considerations ought not to inhibit decisions about developing this part of the hydrologic cycle any more than they do with other portions of the cycle.

3.6 REFERENCES

American Society of Civil Engineers (2004). *Regulated Riparian Model Water Code.* ASCE/EWRI Standard 40-03, Reston, VA, 159 p.

Anderson, F. (1973). *"NEPA in the Courts: A Legal Analysis of the National Environmental Policy Act."* Johns Hopkins University Press, Baltimore, MD.

Beck, R. (1991). *"Augmenting the Available Water Supply,"* R. Beck, (ed.), *Waters and Water Rights,* **1**, (3.04), Michie Company, Charlottesville, VA.

Bluestein, H., Changnon, S., Davis, R., and Orville, H. (1986). "Review and Recommendations on the Use of Safeguards in the North Dakota Weather Modification Program." *Report to North Dakota Weather Modification Board.*

Changnon, S.A., Jr., Davis, R., Farhar, B., Haas, E., Ivens, L., Jones, M., Klein, D., Mann, D., Morgan, G., Sonka, S., Swanson, E., Taylor, R., and Van Blokland, J. (1977). "Hail Suppression: Impacts and Issues." *Illinois State Water Survey,* Champaign, IL.

Colorado Legislative Council (1971). "Controlling Weather Modification Activities." Research Publishers, **174**, Denver, CO.

Davis, R.J. (1968). "The Legal Implications of Atmospheric Water Resources Development and Management." *Report to United States Bureau of Reclamation.*

Davis, R.J. (1974). "Weather Modification Law Developments." *Oklahoma Law Rev.,* **27**, 409.

Davis, R.J. (1975). "Legal Response to Environmental Concerns about Weather Modification." *J. Appl. Meteorology,* **14**, American Meteorological Society, Boston, MA, 681.

Davis, R.J. (1987). "Administrative Inaction on License and Permit Applications." *Report to the Weather Modification Association.*

Davis, R.J. (1988). "The June 1972 Black Hills Flood and the Law." *J. Weather Modification,* **20**, Weather Modification Association, Fresno, CA, 82–87.

Davis, R.J. (1995). "Legal Aspects of Weather Modification Operations." Section 3 of *Guidelines for Cloud Seeding to Augment Precipitation.* ASCE Manuals and Reports on Engineering Practice No. 81 (Kahan et al., eds.), pp. 49–68, ASCE, Reston, VA.

Davis, R.J., and St. Amand, P. (1975). "Proof of Legal Causation in Weather Modification Litigation: Reinbold v. Sumner Farms, Inc., and Irving P. Krick, Incorporated." *J. Weather Modification,* **7**, (1), Weather Modification Association, Fresno, CA, 127–144.

Dewsnup, R., and Jensen, D. (1977). "Legal Aspects of Weather Modification in Utah." Department of Natural Resources, Salt Lake City, UT.

Dobbyn, J. (1989). *Insurance Law in a Nutshell.* West Publishing Company, St. Paul, MN.

Donnan, J., Pellett, J., Leblang, R., and Ritter, L. (1976). "The Rise and Fall of the South Dakota Weather Modification Program." *J. Weather Modification*, 8, (1), Weather Modification Association, Fresno, CA, 1–20.

Gellhorn, E., and Levin, R. (1990). *Administrative Law in a Nutshell.* West Publishing Company, St. Paul, MN.

Getches, D. (1990). *Water Law in a Nutshell.* West Publishing Company, St. Paul, MN, 11–15.

Hurst, W. (1950). *The Growth of American Law: The Law Makers.* Little, Brown and Company, Boston, MA.

Jayson, L. (1984). *Handling Federal Tort Claims: Administrative and Judicial Remedies.* Matthew Bender, NY.

Jones, G. (1991). "Weather Modification: The Continuing Search for Rights and Liabilities." *Brigham Young University Law Rev.*, **1163**.

Keeton, R., Page, W., Dobbs, D., and Owen, D. (1984). *Prosser and Keeton on the Law of Torts.* West Publishing Company, St. Paul MN.

Keyes, C.G., Jr. (2003). "Review of Weather Modification Control Concepts and How to Implement New Mexico Precipitation Enhancement Projects," Draft Report for the NM Interstate Stream Commission, October.

Kionka, E. (1992). *Torts in a Nutshell.* West Publishing Company, St. Paul, MN.

Kirby, J. (1978). "Cloud Seeding in the Texas Panhandle." *J. Weather Modification*, **10**, Weather Modification Association, Fresno, CA, 170–178.

Langerud, D.W. (2004). Communication during review of Section 2, dlangerud@state.nd.us. December.

Mann, D. (1968). "The Yuba City Flood: A Case Study of Weather Modification Litigation." *Bull. Amer. Meteor. Soc.*, **49**, American Meteorological Society, Boston, MA, 690.

New Mexico Law. (2003). Weather Control Act, introduced by Joe M. Stell, 46th Legislature, January.

Pierce, R., Shapiro, S., and Kerkul, P. (1985). *Administrative Law and Process* (7.3–7.4). Foundation Press, Mineola, NY.

Plater, Z., Abrams, R., and Goldfarb, W. (1992). *Environmental Law and Policy: Nature, Law, and Society.* West Publishing Company, St. Paul, MN.

Solak, M.E. (2004). Communication during review of Section 2, msolak@naw-cinc.org. December.

Tarlock, A. (1988). *Law of Water Rights and Resources.* [5.05(4)]. Clark Boardman, NY.

Weather Modification Advisory Board (1978). "Management of Weather Resources." Two Volumes, Report to Secretary, United States Department of Commerce, Washington, DC.

Weather Modification Association (2004a). "Qualifications and Procedures for Certification." *J. Weather Modification*, **36**, Weather Modification Association, Fresno, CA, 105–106.

Weather Modification Association (2004b). "Certified Weather Modification Operators/Managers." *J. Weather Modification*, **36**, Weather Modification Association, Fresno, CA, 107.

Weather Modification Association (2004c). "Statement on Standards and Ethics for Weather Modification Operators." *J. Weather Modification*, **36**, Weather Modification Association, Fresno, CA, 98–104.

SECTION 4
THE SCIENTIFIC BASIS

Robert Czys, Ph. D.[8]*, Thomas P. DeFelice, Ph. D.*[9]*, WMA CO, M. ASCE, and Don A. Griffith*[10]*, WMA CM/CO, A.M. ASCE*

4.1 INTRODUCTION

Precipitation is produced naturally by either collision-coalescence (i.e., gravitationally induced collection of droplets by drops) or as a product of glaciation (i.e., the conversion of liquid to ice and subsequent interaction of ice particles with other ice and/or supercooled liquid), or both of these processes working in tandem. The opportunity for precipitation enhancement arises from the possibility that nucleation agents can be introduced to influence one, the other, or both of these processes to achieve efficiencies greater than what might have occurred naturally. This presumes that at least some clouds are inefficient natural processors of atmospheric water vapor into precipitation.

A strict distinction used to be drawn between seeding to induce primarily a microphysical response, termed "Static-Mode" seeding, and seeding to invigorate cloud motions, termed "Dynamic-Mode" seeding. However, these terms have become somewhat outdated because of a growing awareness that static and dynamic processes work in harmony in nature. It is almost impossible to alter microphysical conditions neither without impacting cloud growth, nor to impact cloud growth without first instilling an effect on the cloud microphysical processes. Hence, it has

[8]Analyst, Department for Research and Development, Communications Data Group, Champaign, IL 61820.

[9]Program Manager, Science & Technology Programs, I.M. Systems, Rockville, MD 20852.

[10]President, North American Weather Consultants, Inc., 8160 S. Highland Dr., Suite A-2, Sandy, UT 84093.

become more common to consider the scientific basis for seeding within the context of seeding technique. There are two seeding techniques in common use: (1) hygroscopic seeding where the collision-coalescence ("warm cloud") process is the target for initial response and (2) glaciogenic seeding, where the "cold cloud" (ice) process is the target for initial response.

This section describes the scientific basis for positive changes in rain. Extensive scientific literature exists on these subjects including scientific reviews by Braham (1986), Dennis (1980), and others listed in the references provided at the end of this section.

4.2 THE NATURAL PRODUCTION OF RAIN

Knowledge about how clouds produce rain naturally is still limited, even though great advances have been made in recent years with regard to instrumentation, the amount and quality of empirical data, theoretical development, and computer simulations of clouds and precipitation. The discussion here is not complete, nor comprehensive, but seeks to provide a basic description adequate to understand how rain enhancement might be achieved.

4.2.1 Formation of Cloud Condensate

The amount of water that can be held in the atmosphere in vapor form is small and dependent on temperature. Air will hold less water in vapor form at colder temperatures than at warmer temperatures. Typically, the atmospheric water vapor actually present at surface temperature and pressure is considerably less than that required for condensation. However, when an air parcel is lifted because of converging winds or some other mechanism, the air expands and cools adiabatically. As this cooling takes place, the absolute amount of water in vapor form remains the same, but the capacity of the air to hold that vapor decreases. Thus, when lifting and cooling is sufficient, a temperature is eventually reached for which the parcel is saturated; i.e., the amount of vapor available equals the carrying capacity of the air. Additional lifting and cooling results in supersaturation; this is when the available water is greater than can be contained in vapor form in the air parcel. At supersaturation the excess water vapor begins to be removed by condensation or deposition in the form of cloud droplets or ice crystals, respectively. It follows then that the amount of cloud condensate in a cloud or cloud system is controlled by (1) the amount of water vapor in an air parcel being lifted,

(2) the amount of lifting, which determines the depth of the cloud, (3) the temperature difference through which condensation takes place, and (4) the extent of the area over which lifting occurs. The cloud condensate formed constitutes the input term for considering precipitation efficiency. If all of this condensate ends up as precipitation reaching the ground, the precipitation efficiency is 100%. If none of it ends up as precipitation reaching the ground, the precipitation efficiency is 0%.

4.2.2 Cloud Initiation and Colloidal Stability

Supersaturation, even that achieved within strong updrafts, usually only exceeds 100% by a few percent. This condition usually is not sufficient to initiate condensation without assistance. The presence of aerosols, such as dusts and pollutants, reduces the supersaturation required for droplet nucleation by providing a surface for the vapor molecules to attach. Hence, the characteristics of the aerosol population ultimately determine the initial cloud droplet concentration and size distribution. The characteristics of the aerosol population are strongly dependent on the underlying surface source regions. For example, continental regions far from oceans are generally associated with high concentrations of small aerosols. In contrast, maritime regions are generally associated with broader aerosol size distributions with smaller concentrations overall. Another factor to consider is that, aerosol concentration tends to increase as surface wind speeds increase.

Cloud condensate over land is nearly always initially available in the form of small droplets, the size and concentration of which are governed by the indigenous aerosol population. This is true even when temperatures in clouds are well below the melting point (0°C)[11]. As a cloud forms, small droplets, typically less than 20 microns in diameter, are generally nucleated in concentrations of hundreds per cubic centimeter. The competition for the water vapor excess among the droplets is strong and further growth by condensation is restricted. Because the fall velocity of these small droplets is slow (<0.3 cm sec^{-1}), they essentially move with the turbulent air currents within the visible cloud and rarely if ever collide with one another. Such clouds are typically referred to as microphysically or colloidally stable because the particles do not interact.

[11] Zero degrees centigrade is the temperature at which all three phases of the pure water substance (vapor, liquid, and ice) coexist in equilibrium and is the threshold above which ice will begin to melt. Liquid drops that exist at temperatures below freezing are referred to as supercooled and the amount of supercooling is given by the temperature deficit below 0°C.

Little or no precipitation that reaches the ground is formed under this condition, and the precipitation efficiency is near 0%.

4.2.3 Initiation and Evolution of Precipitation

Two different mechanisms can disrupt colloidal stability and lead to precipitation measurable at the ground. One mechanism involves direct collision and coalescence among the drops and droplets so that successively larger drops form. When a few large cloud droplets coexist with smaller ones, as determined by the input aerosol population, the larger drops have a slight but significantly greater fall velocity, and grow quickly by the collision-coalescence process. The efficiency with which collisions will occur is highly dependent on the relative sizes and concentrations of the large and small drops. Collisions subsequently become more frequent as the large droplets become progressively larger, due to increasing sweep-out which is the combined effect of increasingly greater cross-sectional area exposed to collisions, and progressively greater fall velocities relative to the smaller droplets. The process accelerates when the drops grow so large that some begin to break up and thereby provide additional embryos for precipitation to evolve. This chain reaction–like process first discussed by Langmuir (1948) leads to progressively greater numbers of precipitation embryos and a growing population of large drops with sufficient fall velocities and mass to reach the ground as precipitation.

The second mechanism that disrupts colloidal stability and may result in precipitation involves the coexistence of supercooled cloud droplets and ice crystals. Because the saturation vapor pressure over ice is less than that over water, ice crystals in the presence of liquid cloud droplets are in an environment that is highly supersaturated with respect to ice (the more numerous water droplets tend to keep the environment at water saturation). Consequently, the ice crystals grow rapidly from vapor transfer to their surface (i.e., vapor deposition). The surrounding water droplets compensate for the vapor loss by evaporating to maintain a saturated vapor environment with respect to water. In turn, the droplets may be replenished by continued nucleation in updraft. Hence, every potential precipitating cloud can be characterized by how its colloidal stability may be disrupted. There are those where (1) collision-coalescence may dominate at temperatures totally warmer than 0°C, (2) the ice process in the presence of supercooled droplets dominates at temperatures totally colder than 0°C, and (3) the process straddles the melting level and collision-coalescence works in tandem with ice processes. This ultimately affects the selection of a seeding technique for precipitation enhancement.

The third mechanism that may disrupt colloidal stability involves direct removal of supercooled cloud droplets by accretion to falling ice crystals. Although the smallest cloud droplets will generally evaporate to compensate for vapor loss to ice crystals growing by diffusion, larger ones survive and can be collected directly by the ice crystals. This rimes the ice crystals and removes substantial quantities of cloud liquid condensate. Intensive riming of crystals leads to graupel and perhaps hail. Riming also facilitates removal of cloud condensate by increasing particle size, mass, and fall velocity, which proportionally increases the sweep-out of droplets and drops. Additional interactions among ice crystals, referred to as aggregation, also promote development of precipitation-size particles which may lead to higher precipitation efficiency.

The precipitation efficiency is not only dependent on the disruption of colloidal stability and creation of hydrometeors that are large enough to achieve fall velocities in excess of the updraft velocity, but also on the fraction of liquid hydrometeor mass that survives evaporation and/or sublimation during its fall toward earth through dry air beneath cloud. The precipitation efficiency can be reduced to zero if the liquid drops or ice particles (graupel or snowflakes) completely evaporate or sublimate beneath cloud to result in virga (precipitation that does not reach the surface). Hence, another aspect to increased precipitation efficiency is when seeding serves to increase mean particle size, concentration, or both such that a greater buffer against evaporation exists to allow a larger fraction of the precipitation mass to reach the earth's surface. Model calculations have also indicated that for equal precipitation-particle mass more of the mass survives the fall in the dry air beneath cloud if the particle starts its descent with a frozen component (Srivastava, 1987).

While collision-coalescence and glaciation processes are the primary focus for reducing colloidal stability, there are interdependencies between mechanisms that should not be ignored. For instance, collision growth can provide very large drops that will freeze more readily because as drop volume increases, the probability that it will contain a natural ice-forming nucleus also increases. It has also been shown that when some larger cloud droplets (>24 μm diameter) co-exist with highly rimed ice particles or graupel in the temperature range from about -3°C to -8°C, accretion can lead to the production of ice splinters (Hallett and Mossop, 1974). The splintering process can enhance ice crystal concentrations by three to four or more factors of 10 above background. Hence, splintering can be an effective mechanism to promote the evolution of precipitation in clouds.

On the other hand, when definite mechanisms for growth of hydrometeors large enough to fall out are not operative, great amounts or even all of the cloud condensate can evaporate before any precipitation is formed; thus, there is very low precipitation efficiency. Clouds that have significant cloud condensate, but do not have appropriate mechanisms for particle growth in the cloud lifetime available, are naturally inefficient and can be considered to have potential for increased precipitation via cloud seeding.

4.3 CLOUD SEEDING TO AUGMENT RAINFALL

There are two seeding techniques in common use: (1) hygroscopic seeding where the warm cloud process is the target for initial response and (2) glaciogenic seeding where the cold cloud process is the target for initial response.

4.3.1 Seeding to Enhance the Warm Cloud Process (Hygroscopic Seeding)

Hygroscopic seeding is generally considered for modifying summer (convective) clouds and has one of two primary objectives: (1) to initiate the warm rain process when it would otherwise not, or (2) to initiate the warm rain process sooner than if not treated. While it is possible to distinguish between many clouds that will initiate the coalescence process and those that will not (Czys and Scott, 1993), it is more difficult to determine when the coalescence process will initiate.

Hygroscopic aerosols and seeding agents are those that have an affinity for water vapor and thus readily allow for the nucleation of water drops at sizes proportional to the initial aerosol size. In the case of hygroscopic seeding, an optimum size for the seeding particle exists such that it will nucleate preferentially over much smaller ambient natural aerosol, grow by condensation, and then continue to preferentially grow by collision-coalescence to precipitation sizes. Recent model simulations indicate that particles around 2 microns in diameter might be optimal (Cooper et al., 1997 and Segal et al., 2004, for example). Recent focus is on producing and delivering optimally sized particles. It should be noted that in some instances model simulations suggest that seeding could result in a net loss of precipitation, even though precipitation was initiated earlier than would have been the case had no seeding occurred. This can be the case because the premature production of precipitation results in "updraft loading" that weakens or may even reverse the updraft that would have promoted continued cloud growth and more overall precipitation (Segal et al., 2004).

When hygroscopic seeding concepts were first introduced the methodology involved either delivering fine sprays of saline solutions, salt ground at upper-cloud levels, or smoke from flares. Hygroscopic seeding did not gain widespread practice in the beginning because each of these delivery mechanisms proved to be impractical. They either involved transport of massive amounts of solutions, generation of particle sizes that could not be adequately controlled and neither could be precisely enough targeted to desired cloud volumes. Furthermore, delivery to cloud top is also undesirable because it only allows for growth to occur over a single downward trajectory.

Recently, "new" hygroscopic seeding flares have been developed that allow for delivery of hygroscopic aerosols at cloud base in size ranges that would positively affect the coalescence process, where the cloud can ingest and then process the aerosol and its subsequent effects along an upward and then downward drop growth trajectory. Mather et al. (1997) discussed the use of such flares in the conduct of a cloud seeding experiment in South Africa. A promising new variation of the use of salt powder for base hygroscopic seeding is provided in Woodley et al. (2005).

Clouds suitable for seeding are those that either would not have an active coalescence process or those that would have delayed initiation of the warm rain process. Seeding usually occurs early in the cloud's lifetime before the warm rain process is firmly established. Hence, clouds suitable for hygroscopic seeding would show strong potential for vertical development, possess moderate updrafts that are strong enough to handle a precipitation load, and would probably have a fairly narrow drop size distribution (but no more so than in the cloud without seeding). Hygroscopic flares would be expended just below cloud base in the area of maximum updraft to ensure that the cloud ingested the seeding agent.

In a way, hygroscopic seeding is scientifically simpler than its glaciogenic counterpart because the primary impact of seeding is to simply promote the coalescence process by either getting it started or by increasing its activity. Hygroscopic seeding may have desirable secondary effects to promote the production of rain. For example, hygroscopic seeding should lead to the production of precipitation-size drops prior to the cloudy air parcel reaching the 0°C isotherm. Consequently, artificially promoted supercooled rain drops may participate in raindrop freezing, to the benefit of latent heat releases, and secondary ice production. These secondary effects are objectives in glaciogenic seeding.

4.3.2 Seeding To Enhance Cold Cloud Process (Glaciogenic Seeding)

The primary purpose of glaciogenic seeding is to initiate the ice process earlier and at a higher level of activity than would occur naturally in order to (1) disrupt colloidal stability by artificially creating ice crystals that may grow by vapor diffusion and/or accretion, (2) invigorate cloud motion through the release of the latent heat of fusion associated with the growth of ice, or both 1 and 2, working in conjunction with one another. The primary active ingredient in the seeding material of choice is silver iodide (AgI) although dry ice is often used. Silver iodide is desirable because it has a molecular structure very similar to ice and thus greatly reduces the excess energy required to create a surface of ice from the vapor and liquid forms. Silver iodide seeding in summertime situations is commonly applied as aerosols created by burning flares dropped into cloud updraft or from ice nuclei generators mounted underneath the airplane's wings dispersed as the airplane flies through or under the cloud.

Clouds suitable for seeding are those that exhibit large amounts of supercooled liquid and low concentrations of ice. They should have at least a weak coalescence process and contain small to moderate concentrations of supercooled drizzle and raindrops to be optimal candidates for seeding, because the nucleated ice particles may grow by riming within the vast reservoir of supercooled cloud water in which they reside. Vertically developing cumulus cloud tops are suitable for seeding shortly after they rise above the 0°C level. Therefore the cloud does not have to be precipitating (internally) before seeding begins. Developing cumulus clouds are especially suitable for seeding if their cloud tops show potential for growing colder than −4°C because AgI is not very active at warmer temperatures. Most of today's glaciogenic seeding agents have some hygroscopic properties and function primarily through the condensation-freezing nucleation mechanism. This being the case, each nucleus forms its own drop, and "selective capture" by larger droplets is not as important as it was a few years ago when AgI functioned primarily by contact-freezing.

The release of latent heat from the freezing of precipitation-size drops is significant but probably overstated by the calculation of Orville and Hubbard (1973), who assumed that the freezing of drops would be instantaneous and isobaric. This is a good assumption for cloud droplets, but becomes increasingly less accurate as drop sizes increase. Laboratory experiments have shown that when a drop freezes, a thin ice shell forms almost immediately (e.g., Schaefer and Cheng, 1971). Concurrently, the release of latent heat raises the mean temperature of the drop to near 0°C. The ice shell then acts as a barrier to the disposal of heat to the environment and thus limits the rate at which the freezing drop converts to solid from the liquid. For larger, millimeter size drops, freezing times can be on the order of 5 to 10 minutes. Therefore, the buoyancy enhancement is nowhere near as dramatic as would be suggested if the drops froze instantaneously and isobarically.

Under circumstances in which glaciogenic seeding material is introduced into cloud volumes where no larger drops are present, seeding serves to primarily disrupt colloidal stability and release additional latent heat which can increase cloud buoyancy. However, when particle growth is restricted to mostly that which can occur by diffusion, the invigorated cloud may "blow" its top and pump condensate into upper troposphere regions reducing precipitation efficiency to nil. Hence, it is important that a cloud character be determined prior to seeding exercises and that seedability criteria be adhered to during operations, so that only clouds having the best chance for positive reaction be selected for treatments.

4.3.3 Seeding to Enhance Development of Individual Convective Clouds

The scientific basis for augmenting precipitation by increasing the precipitation efficiency of clouds has been considered in the previous sections. Precipitation might also be augmented if treatments that disrupt colloidal stability could also lead to larger clouds. Larger clouds have the ability to process more condensate because they often live longer and process greater volumes of water vapor during their lifetimes. This section discusses means by which cloud seeding might enhance cloud development and, consequently, increase precipitation. Orville (1986) reviewed this topic. Other advertent and inadvertent modification methods may well be even more important than cloud seeding for enhancing cloud development, including human-made alterations such as surface albedo changes, urban heat islands, or changes in boundary-layer moisture from large-scale irrigation projects. Each of these alterations that can be made by humans leads, under some circumstances, to changes in boundary-layer heat and moisture fluxes that potentially create significant changes in cloud formation and development and possibly precipitation (Anthes, 1984). Discussion of these factors is beyond the scope of this section; for additional information see Rosenfeld and Woodley (1993).

The basic concept for enhancing the development of individual convective clouds is, in itself, not complicated. Full consideration of the details of the process, however, is quite complex. The simple consideration involves the rapid conversion of large amounts of supercooled cloud water to ice particles. This adds heat to the cloud with respect to the cloud environment, at least in a different place than it would be released by natural ice-nucleating processes, through the release of the latent heat of fusion. In clouds with substantial amounts of supercooled water, this can involve a substantial amount of sensible heat added to the cloud. Cloud model calculations suggest that increases of a few tenths to a degree centigrade or so in appropriate cloud regions can result in modest increases in cloud size. Because the heated cloud air becomes even less dense than the cooler surrounding air, it will become more buoyant and thus rise further than it would have without the additional heating. This can lead to greater (vertical) cloud development. If the air mass in which the convective cloud is embedded is quite stable, the additional heating and buoyancy might have little overall effect. If cases for treatment are selected where the atmosphere is no more than only slightly stable, a slight cloud temperature increase can permit cloud growth to altitudes much higher than those that would be attained by the cloud without the additional heating.

Geographical regions where this type of cloud seeding is feasible are not well defined. Some subtropical areas near the ocean appear to be suitable as well as some continental regions that have access to boundary layer moisture. Clouds that occur in these areas are characterized by warm cloud bases and contain large amounts of supercooled liquid water with appreciable supercooled drizzle and precipitation-size drops. Air masses in such areas can also have mildly stable atmosphere that restrict cloud growth but are not so stable that they preclude deep cloud growth when a small amount of additional heating is added. In contrast, in many continental areas in the mid-latitudes, cloud droplet spectra are much narrower and are composed almost completely of very small cloud droplets. Clouds in these regions also frequently have lower liquid water contents. An additional limitation to this type of cloud seeding in such areas is that, frequently, the thermal stability is not great, and the heat of vaporization released when the cloud droplets are formed is already sufficient to permit cloud growth through the full depth of the troposphere. The effect of the cloud seeding then is to increase the updraft speed, leading to increased condensation rates that may lead to greater precipitation.

4.3.4 Expansion of Glaciogenic Seeding Concepts to Larger Scales

The chain of events hypothesized by Woodley et al. (2003a, 2003b) includes area-wide effects which follow from their Items 6 and 7. The understanding of such cloud seeding enhancement processes is weak due to limited understanding of the linkage between individual cumulus cloud systems and mesoscale processes. Conceptual and numerical models have been used to study how such processes might operate (Fritsch and Chappell, 1981). The various mechanisms that have been proposed require a careful balance between cloud development rates and the movement of the gust front formed from the downdrafts associated with the various cloud towers, as these down drafts reach the surface.

The environmental wind field that is essentially independent of any cloud seeding effect substantially controls the characteristics and movement of the gust front. Thus, expansion of dynamic seeding effects to the mesoscale is likely to be dependent on environmental conditions.

4.4 THE NATURAL PRODUCTION OF SNOWFALL

Most attempts to augment precipitation in the wintertime are focused on increasing snowfall from orographic clouds subject to influence, specifically generation of supercooled liquid water via orographic lift up windward slopes. Winter clouds that occur as part of deep cyclonic storms and mesoscale systems, but do not benefit from orographic influences, have not been the subject of snow augmentation. This has occurred for the most part because additional water to replenish watersheds and increase streamflow is derived from mountainous regions. Therefore, the primary focus of this section is on winter orographic clouds and their potential for precipitation augmentation by seeding.

4.4.1 Formation of Cloud Condensate

The basic physics of the formation and evolution of condensate in winter orographic clouds is basically the same as that for other types of clouds. Lifting and cooling of air to supersaturation results in the nucleation of droplets on the available aerosol to result in cloud production. A primary difference in the process for winter clouds compared to that of summer clouds has to do with the fact that the air is often mechanically lifted over upslope barriers rather than that which results from positive buoyancy. The suspension of cloud droplets initiates as supercooled liquid because cloud formation takes place at temperatures colder than 0°C. Similar to summer clouds, there typically exist very few initial ice crystals because of an absence of natural ice nuclei. Winter clouds are quiescent compared to their summer counterparts; many remain a suspension of supercooled droplets for prolonged periods of time, producing little if any precipitation that reaches the ground.

4.4.2 Cloud Initiation, Colloidal Stability, and Evolution of Precipitation

Initially, winter clouds are usually composed of a population of supercooled droplets characterized by very small mean diameters and a very narrow size range. This, and the absence of ice crystals, makes them extremely colloidally stable, such that it is not uncommon for cloud development near the lee of a mountain range to progress leeward, and eventually evaporate once it detaches from the lifting mechanism to produce little if any snowfall. Other natural clouds might develop with initial populations of supercooled droplets that are less colloidally stable or perhaps initiate ice earlier because of being colder, but are still sub-optimal for precipitation production. Although these clouds may produce some snow at the ground, only a small fraction of the available supercooled water is precipitated. These less efficient clouds are primary candidates for seeding.

4.5 CLOUD SEEDING TO AUGMENT SNOWFALL

The primary purpose of glaciogenic seeding to augment snowfall, as it is for summer clouds, is to convert supercooled liquid to ice and thereby disrupt the colloidal stability of the cloud. As has been pointed out in an earlier section, ice has a depositional growth advantage over liquid at the same temperature and pressure and thus crystals will grow larger (gain more mass) than their liquid counterparts. The initial population of crystals may then follow one of three growth modes, any of which may become more or less important as the precipitation evolves. In the first path, the ice particles may simply grow by deposition, depleting the surrounding population of liquid drops to sizes large enough that they can gain fall speeds larger than the vertical motions within and below the cloud to eventually reach the surface as a fine crystalline snow.

In a second mode, ice crystals grow large enough to gain a differential fall speed relative to the population of droplets; they may then collect droplets through accretion, eventually reaching the surface as heavily rimed grains of ice. This riming process may also result in the production of secondary seeding by rime splintering (Hallett and Mossop, 1974) to further promote the conversion of liquid to ice. In the last, but not the least important, of these growth processes, the ice crystals may attain a sufficiently broadened size distribution that they interact among themselves and aggregate, so that snowflakes eventually reach the surface. For any of these scenarios, the introduction of a seeding agent facilitates ice initiation, allowing depositional growth, riming, and/or aggregation when it otherwise would not occur or would be weaker under natural conditions. Further, a combination of these growth modes may occur over the course of any seeding episode.

4.5.1 Snow Augmentation Methods

Two techniques are commonly used to augment snowfall. Seeding material in the form of silver iodide aerosol is either released directly into suitable clouds using wing tip borne ice nuclei generators or burn-in-place flares on aircraft, or it is released from ground-based generators. Ground-based generators can be either manually operated at lower elevations in populated areas, or operated remotely at higher elevations. Each type of generation system has both advantages and disadvantages, some of which are discussed in Section 5. Dry ice particles can also be dropped by aircraft into regions of clouds containing supercooled liquid water. Although this is a less commonly used material/technique, dry ice seeding from aircraft can offer some specific advantages over silver iodide in terms of targeting and the temperature of the supercooled liquid water that can be treated.

4.5.2 Expansion of Snow Augmentation Concepts to Larger Scales

Because winter orographic clouds often contain significant amounts of supercooled liquid and because seeding is intended to convert much of this liquid to ice, the potential exists for the release of latent heat that may be comparable to that in summer convective clouds. This release of latent heat may either initiate or enhance isolated convective elements within the cloud system, and these elements may act to increase cloud depth. Greater cloud depth may serve to convert more water vapor to supercooled liquid, may provide larger cloud volumes over which precipitation processes may act to produce more snow, and may serve to preserve the cloud for a longer duration than it would have existed naturally. These are all characteristics that promote greater snow amounts at the surface.

Previous research conducted in mountainous regions of the western United States has demonstrated that supercooled liquid water frequently occurs in winter stratiform clouds over the upwind sides of mountain barriers (Warburton and DeFelice, 1986; Reynolds, 1988; Super, 1990; Super, 1999). This supercooled liquid water is often confined to lower elevations in the vicinity of the top of the mountain barrier. Temperatures within these layers can, in some situations, be only slightly colder than 0°C, which presents challenges to effective seeding with glaciogenic seeding agents.

A number of observational and theoretical studies have suggested that there is a cold temperature "window" of microphysical opportunity for cloud seeding. Studies of both orographic and convective clouds have suggested that clouds colder than −25°C have sufficiently large concentrations of natural ice crystals such that seeding can either have no effect or even conceivably reduce precipitation (Grant and Elliott, 1974; Grant, 1986; Gagin and Neumann, 1981; Gagin et al., 1985). It is possible that seeding such cold clouds could reduce precipitation by creating so many ice crystals that they compete for the limited supply of water vapor and result in numerous, more-slowly settling ice crystals that sublimate before reaching the ground. There are also indications that there is a higher temperature limit to seeding effectiveness (Gagin and Neumann, 1981; Grant and Elliott, 1974; Cooper and Lawson, 1984). This is believed to be due to the slow rates of ice crystal vapor deposition growth at warm temperatures and, from a practical standpoint, low efficiency of ice crystal production by silver iodide at temperatures warmer than −4°C. Thus there appears to be a "temperature window" from about −5°C to −25°C at which clouds may respond favorably to silver iodide seeding (i.e., exhibit seedability). Dry ice (frozen carbon dioxide) seeding via aircraft extends this temperature window to temperatures just below 0°C. Liquid propane or other compressed gases, when released directly into supercooled cloud, can initiate ice formation at temperatures similar to those of dry ice, but the release must occur in a cloud.

Orographic clouds are less susceptible to a time window because they are typically quasi–steady state clouds, so they offer a greater time opportunity for successful precipitation enhancement than cumulus clouds. A time window of a different type does exist for orographic clouds, which is related to the time it takes the vapor in a parcel of air to condense to form supercooled liquid water and ascend to the mountain crest. If winds are weak, there may be sufficient time for natural precipitation processes to occur efficiently. Stronger winds may not allow efficient natural precipitation processes, but seeding may speed up precipitation formation. Even stronger winds may not provide enough time for even seeded ice crystals to grow to precipitation sizes before being blown over the mountain crest and sublimating in the sinking subsaturated air to the lee of the mountain. A time window related to the ambient winds, however, is much easier to assess in a field setting for orographic clouds than for cumulus clouds.

If clouds upwind and over mountain barriers containing supercooled liquid water are routinely seeded to produce appropriate concentrations of ice crystals exceeding 10 to 20 per liter of cloudy air, snowfall increases can be anticipated in the presence or absence of natural snowfall. It has been repeatedly demonstrated with physical observations that sufficiently high concentrations of seeding agent, effective at prevailing supercooled cloud temperatures, will produce snowfall when natural snowfall rates are negligible. Seeded snowfall rates are usually light, on the order of 1 mm h^{-1} or less, though consistent with median natural snowfall rates in the intermountain West (Super and Holroyd, 1997).

4.6 TECHNOLOGICAL ADVANCES

Recent advances have been made that improve the chances of making more rapid progress in the field of weather modification and strengthening the scientific basis of the effects of cloud seeding. In the wake of discouraging, inconclusive statistical evaluations of the effects of cloud seeding on precipitation (e.g., Woodley et al., 1983), a new era in cloud measurement (e.g., DeFelice, 1998) and modeling has come on line for all of the atmospheric sciences. An approach complementary to statistical assessment of small precipitation increases, relative to the highly variable natural snowfall rates is to study the basic physics and chemistry involved.

The primary advances regarding instrumentation, which will be discussed further in the next chapter, have been in remote sensing of supercooled liquid water (SLW) and precipitating ice formation, continuous wind and temperature profiling, the development of tracer methods, and the application of improved aircraft mounted sensors. Reviews of the application of several new atmospheric and cloud remote sensing devices to cloud modification and precipitation management are available (Reinking and Meitin, 1989; Reinking, 1992, 1994). The invention of the dual-channel passive microwave radiometer (Hogg et al., 1983) to allow continuous measurements of path-integrated cloud liquid water and vapor has significantly advanced understanding of winter orographic storms and has applications to convective cloud situations as well. Routine use of fixed and scanning radiometers has now produced an abundance of information concerning the availability of SLW to be tapped through judicious application of ice nucleants. Such regions went undetected by aircraft sampling in the past, because of their nearness to the orographic barriers.

Special radars have been developed and further refinements are being made to measure the ice content and graupel and hail development in clouds (Kropfli et al., 1992; Bringi et al., 1989). Computer software has been used with conventional radar to track radar cells and quantify the increased merging due to cloud seeding for dynamic effects (Rosenfeld, 1987; Westcott, 1990). The National Weather Service NEXRAD radars deployed across the United States generate many useful near real-time data displays, including detailed Doppler wind fields on a rapid update (per volume scan) basis. Reinking et al. (1990, 1991, 1992) have used airborne Doppler radar, ozone as a tracer, and numerical cloud models to examine the water budget of a sheared thunderstorm, thus attempting to assess seedability in terms of the storm dynamic structure, its main cell inflow of moisture, losses of ice out the anvil, and rainfall. Additionally, UHF radars that continuously profile the wind (Ecklund et al., 1988) are now commercially available and can be used effectively in project areas to assess airflow and improve targeting of seeding material. Technology transfer from the research to the applications mode is imminent for Radio Acoustic Sounding Systems (RASS) (May et al., 1990) to continuously profile temperature, a measurement that is critical to estimate ice nucleation rate. Reasonably complete measurement of cloud water budgets is now possible (Reinking, 1994).

Tracer methodology has been developed to show where the seeding material goes in a cloud and whether it is effective in nucleating the cloud. Tracers such as sulfur hexafluoride (SF_6) gas (Smith et al., 1992;

Stith et al., 1990; Griffith et al., 1990, 1992) and chaff or aluminized fibers have been used. Chaff is used in association with circular-polarized radar to measure the transport and mixing inside clouds (Martner and Kropfli, 1989; Martner et al., 1992). Indium sesquioxide has been used as a tracer along with silver iodide to determine that silver iodide has participated in forming ice in supercooled orographic clouds (Warburton et al., 1985; Warburton, 1992).

Cloud modeling has improved and computers have become more powerful, with the result that more realistic simulations of convective and stratiform clouds are possible. Cloud seeding tests using computers provide support for the many theories of seeding effects (Orville and Kopp, 1990; Orville, 1990) and a greater understanding of the effects (Orville and Chen, 1982; Orville, 1996; Farley et al., 2000). The strategic use of real-time numerical cloud models combined with continuous, real-time remote sensing offers great potential for predicting and recognizing seeding opportunities and monitoring and determining cloud seeding effects (Reinking, 1994).

4.7 CONCLUSIONS

The scientific basis of precipitation enhancement by cloud seeding has advanced beyond the levels discussed by Grant et al. (1995). In addition, new statistical methods have been designed to aid in the interpretation of cloud seeding operations. However, if further progress is to be made, vigorous use of the available new instrumentation, improved seeding agents, and models must be accelerated. Field efforts need to be designed to take advantage of the available technology, to the extent that it is economically feasible to do so.

4.8 REFERENCES

This section borrows heavily from the version of section 4 that appeared in the first edition of Manual 81 (Grant et al., 1995). The lead author (one listed first) thanks the authors of the 1995 section for providing a firm foundation for this update.

Anthes, R.A. (1984). "Enhancement of Convective Precipitation by Mesoscale Variations in Vegetative Covering in Semi-Arid Regions." *J. Climatology Appl. Meteorology*, **23**, AMS, Boston, MA, 541–554.

Braham, R.R, Jr. (ed.) (1986). "Precipitation Enhancement—A Scientific Challenge." *Meteor. Monogr., 21*, AMS, Boston, MA.

Bringi, N., Sur S., Musil, D., and Rasmussen, R. (1989). "Microphysical Evolution of Convective Clouds Inferred from Multi-Parameter Radar Measurements and Aircraft Penetrations." *Preprints, 24th Conference on Radar Meteorology*, Tallahassee, FL, AMS, Boston, MA, 13–17.

Cooper, W.A., and Lawson, P. (1984). Physical interpretation of results from the HIPEX-1 experiment. *J. Climatology Appl. Meteorology*, **23**, 523–540.

Cooper, W.A., Bruintjes, R.T., and Mather, G.K. (1997). "Some calculations pertaining to hygroscopic seeding with flares." *J. Appl. Meteorology*, **36**, 1449–1469.

Czys, R.R., and Scott, R.W. (1993). "A simple objective method used to forecast convective activity during the 1989 PACE cloud seeding experiment." *J. Appl. Meteorology*, **32**, AMS, Boston, MA, 996–1005.

DeFelice, T.P. (1998). *Introduction to Meteorological Instrumentation and Measurement*, Prentice Hall, Saddle Brook, NJ, 229 p.

Dennis, A.S. (1980). "Weather Modification by Cloud Seeding." *International Geophysics Series*, *24*, Academic Press, New York, NY.

Ecklund, W., Carter, D., and Balsey, B. (1988). "A UHF Wind Profiler for the Boundary Layer: Brief Description and Initial Results." *J. Atmospheric Oceanic Technology*, **5**, AMS, Boston, MA, 432–441.

Farley, R.D., Hjermstad, D.L., and Orville, H.D. (2000). "Numerical simulation of a four-day early spring storm period in the Black Hills." *J. Appl. Meteorology*, **39**, AMS, Boston, MA, 1299–1315.

Fritsch, J.M., and Chappell, C.F. (1981). "Preliminary Numerical Tests of the Modification of Mesoscale Convective Systems." *J. Appl. Meteorology*, **20**, (8), AMS, Boston, MA, 910–921.

Gagin, A., and Neumann, J. (1981). "The second Israeli randomized cloud seeding experiment: evaluation of the results." *J. Appl. Meteorology*, **20**, AMS, Boston, MA, 1301–1311.

Gagin, A., Rosenfeld, D., and Lopez, R.E. (1985). "The relationship between height and precipitation characteristics of summertime convective cells in South Florida." *J. Atmospheric Science*, **42**, AMS, Boston, MA, 84–94.

Grant, L.O., and Elliott, R.D. (1974). "The cloud seeding temperature window." *J. Appl. Meteorology*, **13**, AMS, Boston, MA, 84–94.

Grant, L.O. (1986). "Hypotheses for the Climax wintertime orographic cloud seeding experiments." *Precipitation Enhancement—A Scientific Challenge, Meteor. Monogr.*, **43**, AMS, Boston, MA, 105–108.

Grant, L.O., Orville, H.D., Politovich, M., Reinking, R.F., Rogers D., and Warburton, J. (1995). "The Scientific Basis," ASCE Manual of Professional Practice No. 81, Section 4, *Guidelines for Cloud Seeding to Augment Precipitation*, ASCE, Reston, VA, 69–85.

Griffith, D.A., Wilkerson, G.W., and Risch, D.A. (1990). "Airborne Observations of a Summertime, Ground-Based Tracer Gas Release." *J. Weather Modification*, **22**, WMA, Fresno, CA, 43–48.

Griffith, D.A., Wilkerson, G.W., Hauswe, W.J., and Risch, D.A. (1992). "Observations of Ground Released Sulfur Hexafluoride Tracer Gas Plumes in Two Utah Winter Storms." *J. Weather Modification*, 24, WMA, Fresno, CA, 49–65.

Hallett, J., and Mossop, S.C. (1974). "Production of secondary ice crystals during the riming process." *Nature*, **249**, 26–28.

Hogg, D.C., Guiraud, F.O., Snider, J.B., Decker, M.T., and Westwater, E.W. (1983). "A Steerable Dual-Channel Microwave Radiometer for Measurement of Water Vapor and Liquid in the Troposphere." *J. Appl. Meteorology*, **22**, AMS, Boston, MA, 789–806.

Kropfli, R., Matrosov, S., Frisch, S., and Orr, B. (1992). "The Use of Radar in Cloud/Radiation Studies: Initial Results from Claret 1, Fire Ii, and Astex." *Workshop on Cloud Microphysics and Applications to Climate Change*, 10–14 August 1992, Toronto, Canada, World Meteorological Organization.

Langmuir, I. (1948). "The production of rain by a chain reaction in cumulus clouds at temperatures above freezing." *J. Meteorology*, **5**, AMS, Boston, MA, 175–192.

Martner, B.E., and Kropfli, R.A (1989). "Tracir: A Radar Technique for Observing the Exchange of Air Between Clouds and Their Environment." *Atmospheric Environment* **23**, AMS, Boston, MA, 2715–2721.

Martner, B.E., Marwitz, J.D., and Kropfli, R.A. (1992). "Radar Observations of Transport and Diffusion in Clouds and Precipitation Using Tracir." *J. Atmospheric Oceanic Technology*, **9**, AMS, Boston, MA, 226–241.

Mather, G.K., Terblanche, D.E., Steffens, F.E., and Fletcher, L. (1997). "Results of the South African cloud-seeding experiments using hygroscopic flares." *J. Appl. Meteorology*, **36**, AMS, Boston, MA, 1433–1447.

May, P., Strauch, R., Moran, K., and Eklund, W. (1990). "Temperature Soundings by Rass with Wind Profiler Radars." *IEEE Transcripts of Geoscience Remote Sensing*, **28**, 19–28.

Orville, H.D. (1986). "A Review of Dynamic Mode Seeding of Summer Cumuli." From Precipitation Enhancement—a Scientific Challenge, R. Braham, (ed.), *Meteor. Monogr.*, **21**, (43), AMS, Boston, MA, 43–62.

Orville, H.D. (1990). "The Uses of Cloud Models in Weather Modification." *J. Weather Modification*, **22**, WMA, Fresno, CA, 137–142.

Orville, H.D. (1996). A review of cloud modeling in weather modification. *Bull. Amer. Meteor. Soc.*, **77**, AMS, Boston, MA, 1535–1555.

Orville, H.D., and Chen, J.M. (1982). "Effects of Cloud Seeding, Latent Heat of Fusion and Condensate Loading on Cloud Dynamics and Precipitation Evolution: A Numerical Study." *J. Atmospheric Sciences*, **39**, AMS, Boston, MA, 2807–2827.

Orville, H.D., and Hubbard, K. (1973). "On the freezing of liquid water in a cloud." *J. Appl. Meteorology*, **12**, AMS, Boston, MA, 671–676.

Orville, H.D., and Kopp, F.J. (1990). "A Numerical Simulation of the 22 July 1979 Hiplex-1 Case." *J. Appl. Meteorology*, **29**, AMS, Boston, MA, 539–550.

Reinking, R.F. (1992). "The NOAA Federal/State Cooperative Program in Atmospheric Modification Research: A New Era in Science Responsive to Regional and National Water Resources Issues." *Preprints, Symposium on Planned and Inadvertent Weather Modification*, Atlanta, GA, AMS, Boston, MA, 136–144.

Reinking, R.F. (1994). "An Approach to Remote Sensing and Numerical Modeling of Orographic Clouds and Precipitation for Climatic Water Resources Assessment." *Atmospheric Research*, **30**, AMS, Boston, MA.

Reinking, R.F., and Meitin, R.J. (1989). "Recent Progress and Needs in Obtaining Physical Evidence for Weather Modification Potentials and Effects." *J. Weather Modification*, **21**, WMA, Fresno, CA, 85–93.

Reinking, R.F., and Colleagues. (1991). "The Lake Ontario Winter Storms (Lows) Project." NOAA T.M. ERL WPL-216 (USGPO: 1991-673002/41,024).

Reinking, R.F., Stith, J.L., and Meitin, R.J. (1990). "Airborne Doppler Radar and in Situ Studies of the Transport of Ozone and Other Constituents in Feeder Cells and Anvils." *Preprints, Conference on Cloud Physics*, San Francisco, CA, AMS, Boston, MA, 698–705.

Reinking, R.F., Meitin, R.J., Kopp, F., and Orville, H.D. (1992). "Fields of Motion and Transport within a Sheared Thunderstorm." *Atmospheric Research*, **28**, 197–226.

Reynolds, D. (1988). "A Report on Winter Snowpack Augmentation." *Bull. Amer. Meteor. Soc.*, **69**, AMS, Boston, MA, 1290–1300.

Rosenfeld, D. (1987). "Objective Method for Tracking and Analysis of Convective Cells as Seen by Radar." *J. Atmospheric Sciences*, **4**, AMS, Boston, MA, 422–434.

Rosenfeld, D., and Woodley, W.L. (1993). "Effects of Cloud Seeding in West Texas: Additional results and new insights." *J. Appl. Meteorology*, **32**, AMS, Boston, MA, 1848–1866.

Schaefer, V.J., and Cheng, R.J. (1971). "The production of ice crystal fragments by sublimation and electrification." *J. Rech Atmo*, **5**, 5–10.

Segal, Y., Khain, A., Pinsky, M., and Rosenfeld, D. (2004). "Effects of hygroscopic seeding on raindrop formation as seen from simulations using a 2000-bin spectral cloud parcel model." *Atmospheric Research*, 71, 3–34.

Smith, P.L., Orville, H.D., Boe B.A., and Stith, J.L. (1992). "Weather Modification in the Dakotas." *Atmospheric Research*, **28**, 271–298.

Srivastava, R.C. (1987). "A model of intense downdrafts driven by the melting and evaporation of precipitation." *J. Atmospheric Sciences*, **44**, AMS, Boston, MA, 1752–1773.

Stith, J.L., Detwiler, A.G., Reinking, R.F., and Smith, P.L (1990). "Ingestion, Transport, Mixing and the Formation of Ice in Cumuli with Gaseous Tracer Techniques." *Atmospheric Research*, **25**, AMS, Boston, MA, 195–216.

Super, A.B. (1990). "Winter Orographc Cloud Seeding Status in the Intermountain-west." *J. Weather Modification*, **22**, 106–116.

Super, A.B., and Holroyd, E.W. (1997). "Some physical evidence of AgI and liquid propane seeding effects on Utah's Wasatch plateau." *J. Weather Modification*, **29**, 8–32.

Super, A.B. (1999). "Summary of the Utah/NOAA Atmospheric Modification Program: 1990–1998." *J. Weather Modification*, **31**, 51–75.

Warburton, J. (1992). "Atmospheric (Weather) Modification in the Decade of the 80's." *Preprints, Symposium on Planned and Inadvertent Weather Modification*, Atlanta, GA, AMS, Boston, MA, 5–11.

Warburton, J., and DeFelice, T.P. (1986). "Oxygen isotopic composition of central Sierra Nevada precipitation. I. Identification of ice phase water capture regions in winter storms." *Atmospheric Research*, **20**, 11–22.

Warburton, J., Young, L., Owens, M., and Stone, R. (1985). "The Capture of Ice Nucleating and Non-Ice Nucleating Aerosols by Ice Phase Precipitation." *J. Rech. Atmo.*, **19**, 249–255

Westcott, N.E. (1990). "Radar Results of the 1986 Exploratory Field Program Relating to the Design and Evaluation of Pace." *J. Weather Modification*, **22**, WMA, Fresno, CA, 1–17.

Woodley, W.L., and Rosenfeld, D. (1992). "Seeding Supercooled Cumuli for Rain Enhancement; Progress and Prospects." *Preprints, Symposium on Planned and Inadvertent Weather Modification*, Atlanta, GA, AMS, Boston, MA, 18–27.

Woodley, W.L., Barnston, A.G., Flueck J.A., and Biondini, R. (1983). "The Florida Area Cumulus Experiment's Second Phase (Face-2). Part II: Replicated and Confirmatory Analyses." *J. Climate Appl. Meteorology*, **22**, AMS, Boston, MA, 1529–1540.

Woodley, W.L., Rosenfeld, D., and Silverman, B.A. (2003a). "Results of on-top glaciogenic cloud seeding in Thailand: part I. The demonstration experiment." *J. Appl. Meteorology*, **42**, AMS, Boston, MA, 920–938.

Woodley, W.L., Rosenfeld, D., and Silverman, B.A. (2003b). "Results of on-top glaciogenic cloud seeding in Thailand: part II. Exploratory analyses." *J. Appl. Meteorology*, **42**, 939–951.

Woodley, W.L., Rosenfeld D., Axisa D., Lahav R., and Bomar, G. (2005). "On the Documentation of Microphysical Signatures Following the Base-Seeding of Convective Texas Clouds Using Salt Micro-Powder." 16th Conf. on Wea. Mod. AMS, Boston, MA.

SECTION 5

CLOUD SEEDING MODES, INSTRUMENTATION, AND STATUS OF PRECIPITATION ENHANCEMENT TECHNOLOGY

Don A. Griffith[12], *A.M. ASCE, WMA CM/CO*

5.1 INTRODUCTION

Once the decision has been made to implement a precipitation augmentation cloud seeding project or program, consideration needs to be given to the project or program design. Such a design is needed in order to systematically consider the important aspects of setting, conducting, and evaluating a project. The design should include: (1) a statement of the goals of the project; (2) definition of the project area (target and adjacent affected area) and control area, if any; (3) selection of an operational period; (4) specification of cloud seeding modes; (5) project instrumentation requirements; (6) cloud seeding criteria; (7) suspension criteria; and (8) evaluation criteria. Such a design can provide an excellent source of information for the preparation of a solicitation requesting the work to be performed. Several aspects relating to a project design are examined elsewhere in this report. The intent of this section is to provide a description of the possible cloud seeding modes and instrumentation for use in conducting cloud seeding operations. These two topics are covered separately in the following sections.

5.2 CLOUD SEEDING MODES

"Cloud seeding modes" is a term utilized to denote the choices available in selecting the appropriate cloud seeding agent, as well as the methods available for dispensing these agents. There are several decisions to

[12]President, North American Weather Consultants, 8160 South Highland Dr., Ste. A-2, Sandy, UT 84093

be made in determining the best cloud seeding mode for a particular project. What cloud seeding agent to use—dry ice, silver iodide, liquid propane, or some organic compound? Should ground generators or airborne generators or a combination of the two be used? At what rate should the seeding agent be applied? How many ground generators or seeding aircraft are required to adequately seed the specified target area? Decisions such as these are best made on the basis of recommendations of knowledgeable meteorologists working in the field of weather modification. Sometimes the most desirable cloud seeding mode will prove to be too costly when compared with probable economic benefits, which will necessitate the specification of a more economical but still reliable cloud seeding methodology (Griffith et al., 1995).

5.2.1 Cloud Seeding Agents

Experiments conducted at the General Electric Laboratories in Schenectady, New York, in 1946 and 1947 by Drs. Schaefer (1946) and Vonnegut (1947) demonstrated that certain materials are quite effective in converting supercooled liquid water droplets (droplets at temperatures lower than 0°C) into ice crystals. Schaefer demonstrated that dry ice (solid carbon dioxide) particles, when dropped through a cloud, produce ice crystals due to spontaneous nucleation (Mossop, 1955). Additional experiments conducted by Vonnegut were concerned with the possible identification of materials that might serve to promote heterogeneous nucleation (Griffith et al., 1995). Heterogeneous nucleation, which is by far the dominant process in nature, can result either from direct deposition to ice from the vapor phase or the direct contact nucleation of supercooled liquid water. Foreign aerosol particles can promote both processes. In the case of supercooled liquid water, such aerosol particles can significantly raise the temperature at which the drops would otherwise freeze by homogeneous nucleation. Among the most effective freezing nuclei identified by Vonnegut were silver iodide and lead iodide (AgI and PbI). The temperature threshold at which particles of these substances began to produce a few ice crystals was in the range of −3°C to −4°C, much higher than with most naturally occurring substances in the atmosphere.

Later research (Fukuta, 1963, 1966) demonstrated that several organic materials can also provide effective freezing nuclei. Two examples of such materials are 1,5-dihydroxynaphtalene and metaldehyde. *Pseudomonas syringae* (Ward and Demott, 1989), a bacteria thought to reduce frost damage in plants, has also been shown to be an effective heterogeneous nucleating agent. Different classes of cloud seeding agents (homogeneous, heterogeneous, and organic, either homogeneous or heterogeneous) will be examined separately. A fourth class of cloud seeding agents, which consists of an array of hygroscopic (water absorbing) materials, will also be covered.

5.2.1.1 Homogeneous Nucleating Agents. Dry ice (solid CO_2) is an effective ice-nucleating agent producing 2×10^{11} to 8×10^{11} ice crystals per gram of dry ice dispensed, and its effectiveness is relatively independent of temperature in the range from $-1°C$ to $-11°C$ (Holroyd et al., 1978). Other research, (e.g., Horn et al., 1983) indicates that these numbers may be several orders of magnitude low, with definite temperature dependence. While dry ice has a number of advantages, such as rapid transformation of supercooled cloud water droplets and vapor to ice, good effect near the melting level, a total lack of any toxic residuals, and low cost, it must be dispensed directly into the supercooled region of the cloud—thus usually requiring an airborne delivery system. Dry ice was frequently used in cloud seeding projects in the United States in the 1950s and early 1960s but was slowly replaced by silver iodide as convenient storage and dispensing capabilities for generation of silver iodide particles were developed. Dry ice has received some attention again in recent years, especially for research projects.

Liquid propane is a freezing agent much like dry ice. It produces almost the same number of crystals per gram as does CO_2 (Kumai, 1982). It cannot be dispensed from aircraft because it is a flammable substance. However, it can be dispensed from the ground if released at elevations which are frequently within supercooled clouds. The U.S. Air Force has used liquid propane dispensed from ground-based sites to clear supercooled fog at military airports for more than 30 years. It has recently been applied as a cloud seeding agent for winter snowpack enhancement through development of a remotely operated ground-based dispenser (Reynolds, 1991, 1992). Liquid propane seeding experiments were also conducted on the Utah/National Oceanic & Atmospheric Administration's Atmospheric Modification Project (Super, 1999). Future experimentation needs to demonstrate that this technique can increase precipitation over a fixed target area for a significant period of time (e.g., a winter season).

5.2.1.2 Heterogeneous Cloud Seeding Agents. Both silver iodide (AgI) and lead iodide (PbI) particles, in size ranges of 0.1 micron to 1 micron are very effective freezing nuclei. Due to environmental concerns related to the release of lead into the atmosphere, PbI has generally not been utilized as a cloud seeding agent. However, silver iodide does not suffer from these environmental constraints and has been the preferred seeding agent in many cloud seeding projects. This has been the case for projects and/or programs conducted in the United States and in other countries, such as Australia and Canada, for decades. Correctly sized particles of AgI are usually produced through some combustion process followed by rapid quenching, which forms literally billions of effective freezing nuclei/gram of AgI consumed if temperatures are lower than $-4°C$. Methods of generating properly sized particles of AgI are covered in Section 5.2.2.

Cloud chambers have been constructed at several research facilities, such as Colorado State University (1969–2004), in order to test, among other things, the effectiveness of different generating techniques in producing freezing nuclei. Figure 5-1 provides plots of the number of effective freezing nuclei produced per gram of AgI consumed in a variety of seeding material generation methods as a function of the temperature of the cloud chamber (Garvey, 1975). All of the curves in Figure 5-1 exhibit an increase in activity with decreasing temperatures. Similar studies of naturally occurring freezing nuclei exhibit this same characteristic with even fewer active nuclei in the regions warmer than −10°C. Note the relative increase in effectiveness of silver iodide-ammonium iodide mixtures over that of silver iodide and sodium iodide. The presence of either ammonium or sodium iodide in solutions of AgI and acetone mixtures is prompted by the need for a catalyst to allow dissolving AgI in acetone. The differences in activity from solutions involving sodium and ammonium iodide have been studied extensively and are attributed to more hygroscopic complexes being formed using sodium iodide solutions. This decreases the effectiveness of AgI particles produced from these solutions (Donnan et al., 1970). Following this research, conducted by the Naval Weapons Center, the South Dakota School of Mines and Technology, and other groups, most operations utilizing AgI acetone solutions in that era (1970s) switched from sodium iodide to ammonium iodide as the catalyst. Differences also occur due to the difference in airflow past the cloud seeding generator. This effect is shown in the upper left hand figure, where maximum fan production is greater, and this might be expected to occur under cloud seeding conditions.

Later research by DeMott et al. (1983) in cloud chamber tests indicates that the addition of ammonium perchlorate (NH_4ClO_4) to the standard silver iodide, ammonium iodide, acetone solution increases the number of effective freezing nuclei produced per gram of silver iodide. Other research (DeMott, 1988) indicates that the addition of sodium perchlorate ($NaClO_4$) to the above solution apparently produces a hygroscopic nucleating agent. The complex produced through the burning of these mixtures exhibits significantly faster reaction times in producing ice crystals than other means of producing AgI freezing nuclei. Finnegan (1999) provides further documentation that the addition of chlorine to seeding solutions provides increased activity at higher temperatures and, perhaps more important, faster rates of nucleation than solutions that generate pure AgI. These results suggest that there are choices available in the design of cloud seeding projects that utilize AgI in terms of the potential number of ice crystals that can be produced per gram of AgI utilized and also in the rates at which these ice crystals are produced. Different cloud seeding project designs may be able to effectively utilize these choices to a definite advantage.

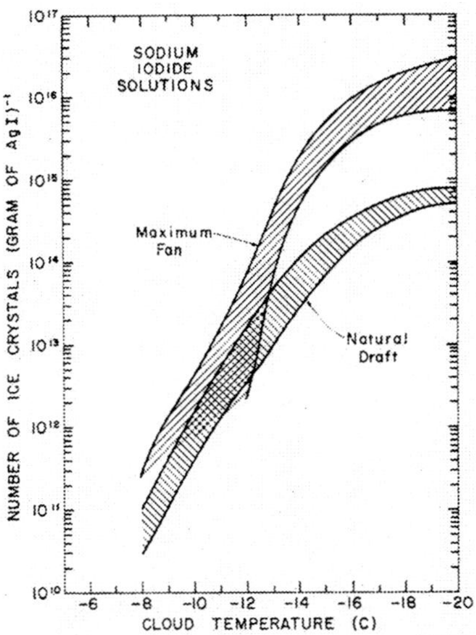

Ranges of effectiveness values for the aerosols produced by five ground generators burning solutions of AgI-NaI in acetone for two conditions of ventilation.

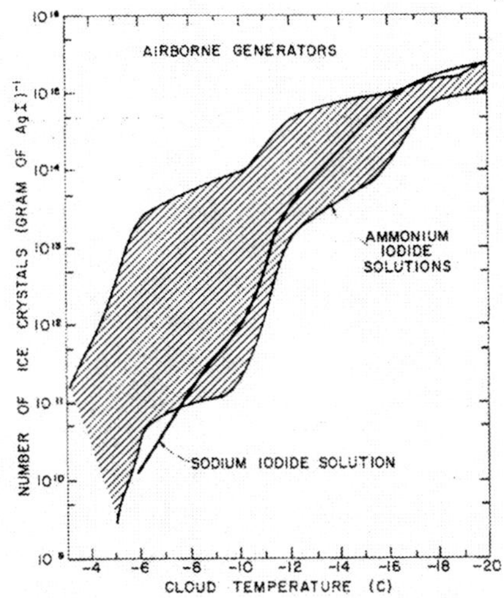

Range of effectiveness values for the aerosols produced by three airborne generators burning various solutions of AgI-NH_4I and one solution of AgI-NaI.

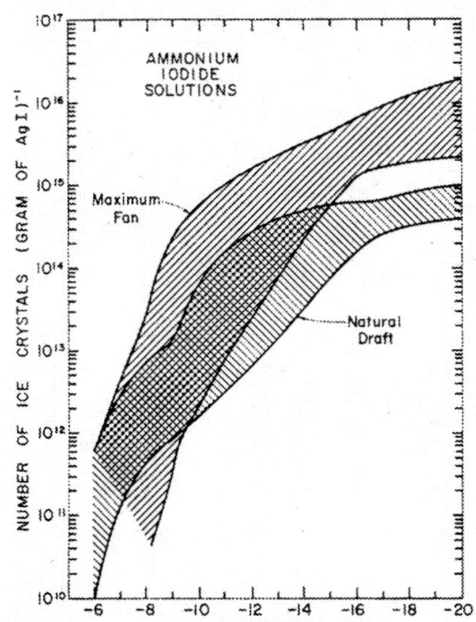

Ranges of effectiveness values for the aerosols produced by six ground generators burning solutions of AgI-NH_4I in acetone for two conditions of ventilation.

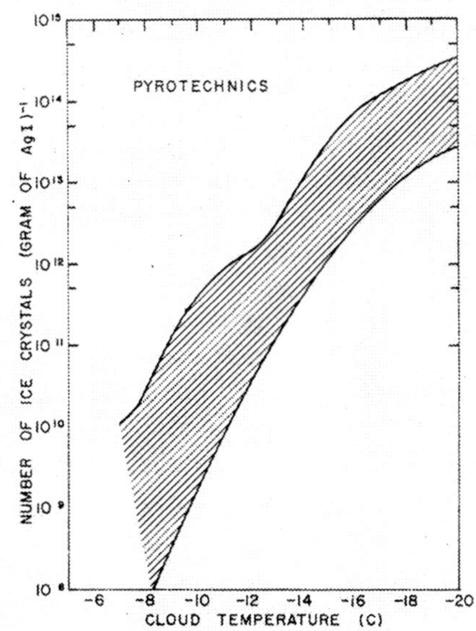

Range of effectiveness values for the aerosols produced by ten kinds of pyrotechnics.

FIGURE 5-1. Ranges of Effectiveness Values for Aerosols Produced by Various Methods (Griffith et al., 1995).

It has been shown that field observations of nucleation efficiency from wingtip acetone generators or ground-based silver iodide generators may exhibit even higher efficiencies (Finnegan and Pitter, 1987). This is due to the production of water vapor in the combustion process, providing transient supersaturations. Deshler et al. (1990) report on such a phenomenon from aerial seeding in clouds over the Sierra Nevada.

The potential toxicity of AgI and possible environmental impacts have been investigated extensively (Cooper and Jolly, 1969; Douglas, 1968; Klein, 1978). The results of this research indicate little concern over the short term (measured in decades), with some possible impact in the long term (hundreds of years).

5.2.1.3 Organic Cloud Seeding Materials. Several organic materials (compounds of carbon) have been identified as effective freezing nuclei. Phloroglucinol appears to have been the first organic freezing nucleant identified (Langer et al., 1963). Several other compounds, including trichlorobenzene, raffinose, trimesic acid, melamine, 1-leucine, 1-tryptophane, metaldehyde, and 1,5-dihydroxynaphtalene have been shown to provide effective freezing nuclei (Fukuta, 1963 and 1966; Langer et al., 1963; Power and Power, 1962). Some of these compounds exhibit higher threshold activation temperatures, which hold promise for potential future applications where activation near the melting level is advantageous. Most organic nucleants have received little field-testing to date; therefore, their acceptance for use in operational precipitation enhancement projects will probably be delayed until tested in research-oriented cloud seeding projects. The exception is liquid propane, which has been previously mentioned. Its efficiency is similar to that of solid carbon dioxide (Griffith et al., 1995).

There are several potential advantages associated with organic nucleants, which should encourage consideration for testing in research projects. These many materials are biodegradable, have lower costs (especially when compared to silver iodide), and have comparatively high temperature activation thresholds. Some organic materials, unlike dry ice, can undoubtedly be adapted to ground generation techniques (similar to propane).

5.2.1.4 Hygroscopic Materials. Numerous precipitation enhancement projects have been using AgI complexes as their primary nucleating agent since the 1950s (ASCE, 2004). Nevertheless, the injection of hygroscopic agents which may alter the initial cloud droplet spectra or create raindrop embryos immediately may be an efficient method for treating warm-based continental cumulus clouds, in which the vertical distance from cloud base to the melting level can be as much as a few kilometers. Ludlam (1958) and Appleman (1958) described the concepts involved in hygroscopic seeding with salt particles by dropping large numbers of salt particles into cumulus clouds. Salt seeding was used experimentally in the North Dakota Pilot Project, a combination hail suppression and rainfall enhancement project, in 1972. In that experiment and others conducted in South Dakota, finely ground salt particles were released near the bases of moderate sized cumulus clouds to create raindrop embryos around the salt particles. Experiments carried out in South Africa in the early 1990s underscored the potential effectiveness of seeding with hygroscopic agents in increasing precipitation from cumulus clouds (Mather and Terblanche, 1994).

Hygroscopic agents deliquesce (that is, become liquid by absorbing moisture from the air) at relative humidity values significantly less than 100%. Mather et al. (1997) made use of flares containing primarily potassium perchlorate, which when burned produced potassium chloride (KCl) particles with mean diameters of about 0.5 microns. The hygroscopic flares contained about 1 kilogram of seeding material. These flares were burned near the base of cumulus clouds in an attempt to alter the cloud droplet spectra through the "competition" effect. Although there are many naturally occurring hygroscopic substances, KCl particles have an advantage of only requiring a relative humidity on the order of 70% to 80% to deliquesce, and readily act efficiently as CCN.

Project planners should bear in mind that the hygroscopic flare method is relatively new and is not yet used as widely as the AgI complexes, although it has shown considerable promise (Cooper et al., 1997; Mather et al., 1996, 1997). Additional experimentation utilizing this technique has been conducted in Mexico for rain enhancement (Bruintjes et al., 1999). Future experimentation needs to demonstrate that this technique can increase precipitation over a fixed target area for a significant period of time (e.g., a summer convective season).

5.2.1.5 Other Seeding Methods and Inadvertent Weather Modification. Over the years, there have been a number of proposed techniques to produce increases in precipitation. Some of these techniques are based upon plausible physical principles, and may offer potential (though they are yet to be proven) through field-testing. An example could be the alteration of the albedo of an area through the installation of a surface that absorbs the sun's energy, thereby creating increased convection near the earth's surface, possibly leading to enhanced cloud development. An inadvertent effect appears to have been detected during the Metromex project (Changnon et al., 1971) conducted in the St. Louis area, although, in this case, the effect was hypothesized to have been due to urban pollution or heat island effects. Recent research (Rosenfeld and Lensky, 1998) has indicated that "natural and anthropogenic aerosols can substantially modify clouds not only in pristine environments, as was already demonstrated by the ship tracks, but they can also produce a major impact on cloud microstructure and precipitation in more continental environments, leading to substantial weather modification in densely populated areas."

Substantial reduction in the rainfall efficiency of clouds was observed to be induced by plumes of smoke caused by biomass burning due to agricultural practices, forest fires (Rosenfeld, 1999; Andreae et al., 2004), cooking and heating, and industrial processes (Rosenfeld, 2000; Ramanathan et al., 2001). This was manifested as actual loss of 15% to 25% of the winter precipitation from orographic clouds downwind of coastal major urban areas (Givati and Rosenfeld, 2004). Air pollution was observed to mask the added rainfall due to cloud seeding (Givati and Rosenfeld, 2005). Therefore the opposing effects of air pollution and cloud seeding need to be separated for proper assessment of the anthropogenic impacts on precipitation amounts.

Polluted clouds with suppressed precipitation could regain their raining ability once they incorporate large hygroscopic particles originating as sea spray (Rosenfeld et al., 2002) or salt dust particles from salt flats such as the anthropogenically dried Aral Sea (Rudich et al., 2002) because large hygroscopic particles were then mixed into the clouds and overrode the detrimental effect of the smoke particles.

5.2.2 Delivery Systems

A number of alternatives exist regarding cloud seeding delivery systems. A basic division exists between these alternatives, consisting of ground-based or aerial generating systems. Most systems currently in use are designed to dispense either silver iodide nuclei or particles of dry ice. The choice of the delivery system (or systems) should be made on the basis of the project design, which should establish the best system for the specific requirements and the topographic configuration of a given project (Griffith et al., 1995).

5.2.2.1 Aerial Application. Commonly available aircraft can be modified to carry an assortment of cloud seeding devices. Silver iodide nuclei dispensers include models which burn a solution of silver iodide dissolved in acetone, or through pyrotechnic devices (flares), either droppable or burn-in-place units. A typical silver iodide solution burner has a solution tank and a nozzle configuration. The silver iodide acetone solution is forced through the nozzle into a combustion chamber where the atomized solution is ignited, and the silver iodide crystals formed through combustion are expelled along with the other combustion by-products into the atmosphere (Griffith et al., 1995; ASCE, 2004).

Pyrotechnics are similar to ordinary highway flares that are typically ignited at one end and designed to burn for varying periods of time from several seconds to several minutes. Cloud seeding pyrotechnics (often referred to as flares) are impregnated with varying amounts of silver iodate ($AgIO_3$); $AgIO_3$ is used since this compound provides the oxygen needed to burn the flare formulation. They are classified as Class 1.4 pyrotechnics, which require some restrictions in the way they are transported. Cloud seeding pyrotechnics can be burned from racks mounted on an aircraft near the trailing edge of the wing or can be dropped from the underside of the aircraft. In the latter case, the flare is ignited as it leaves the aircraft and then falls for approximately 600 to 1,800 m (depending upon the designed burn time) before being completely consumed. An aluminum casing containing the droppable pyrotechnic mixture remains in the dispensing rack on the aircraft when the cloud seeding mixture is expelled by a propellant charge. Pyrotechnics typically produce 10 to 100 grams of active seeding agent per minute of AgI, while aerial acetone generators typically produce 2 to 3 grams of active seeding agent per minute of AgI (ASCE, 2004). The rate at which the seeding agent is dispersed is not the only important factor, however. Cloud chamber tests indicate that, in general, acetone generators produce about ten times as many effective ice nuclei per gram of AgI burned as do pyrotechnics. In addition, the activation temperatures and nucleation mechanisms may also vary. All of these factors should be considered when selecting the type of generation method to be used. Laboratory cloud chamber test results can be very informative in this regard. Figures 5-2 through 5-4 provide common installations of an acetone dispenser, a pyrotechnic burn-in-place rack, and a rack for droppable pyrotechnics.

FIGURE 5-2. Acetone-Silver Iodide Generator Mounted on a Wing Tip (ASCE, 2004).

FIGURE 5-3. Wing-Mount Silver Iodide Pyrotechnic Rack (ASCE, 2004).

FIGURE 5-4. Droppable Silver Iodide Pyrotechnic Rack (ASCE, 2004).

Dry ice is frequently dispensed through openings in the floor of baggage compartments or extra passenger seat locations on modified cloud seeding aircraft. Dispensers have been designed to disperse "pelletized" or small particles of dry ice. Dry ice pellets are available commercially in some of the larger cities of the United States. Diameters of 0.6 to 1 cm and 0.6 to 2.5 cm in length are the appropriate size. The goal of dispensing dry ice is to have the particles fall 1 to 2 km before they sublime completely, thereby creating a sizable "curtain" of seeded cloud area. Other dispensers have been developed that either dispense pre-crushed dry ice or actually crush dry ice slabs onboard the aircraft. Figure 5-5 provides a photograph of a dry ice crusher mounted in an aircraft. A third type of dispenser developed for Air Force research in the 1960s (Vickers and Church, 1966) manufactured dry ice pellets from containers of pressurized liquid carbon dioxide.

Some prototype organic and hygroscopic dispensers have been developed on various projects. Fukuta et al. (1977) has reported on an organic dispenser that received some field-testing in South Dakota. Some agricultural spray dispensers have been modified to dispense hygroscopic materials. One disadvantage of most hygroscopic materials is that they are corrosive, requiring special care in order to avoid damage to the cloud seeding aircraft. They also tend to clump in humid conditions, requiring careful storage and handling techniques.

Types of aircraft utilized in operational cloud seeding projects range from an occasional single-engine aircraft (such as a Cessna 182, or Piper Comanche) to larger twin-engine aircraft (Piper Twin Comanche, Aztec, Navajo, Seneca, and Cheyenne, or Cessna 310, 340, 411, 414, and 421, or Aerocommander 690)(ASCE, 2004). Any modification of an aircraft incorporating cloud seeding equipment must be certified by the Federal Aviation Administration (FAA) and this certification usually places them in a restricted category. As the name implies, there are certain restrictions governing the use of such aircraft, including a limitation on the type of personnel authorized to fly in such aircraft.

FIGURE 5-5. Dry Ice Dispenser Mounted in an Aircraft (ASCE, 2004).

It is important to remember that the type of cloud seeding agent and delivery system used may dictate the type of aircraft that can be used. Dry ice or droppable AgI flares are usually dispensed at cloud top. However, this is only possible if cloud tops are accessible to seeding aircraft, are tall enough to contain supercooled water (0°C to −10°C), and are positioned such that proper targeting of the cloud seeding effects is possible. Cloud seeding flights that are likely to entail flying through supercooled portions of clouds should be conducted with aircraft with deicing capabilities.

For AgI acetone burners and end burning flares, the aircraft can be operated in updraft regions below cloud base. However, it is advantageous to directly inject the seeding agent into supercooled cloud. With possible flight durations of four hours or more, the aircraft should be fully deiceable or frequent descents below the melting level will be required to shed ice buildup.

Recent research in winter orographic cloud seeding utilizing aircraft suggests that AgI acetone wingtip generators provide the simplest and most effective way to seed from aircraft. This, because (1) a 30-liter solution tank holds enough cloud seeding solution for a 5-hour flight; (2) silver iodide solutions with a perchlorate additive can effectively seed at temperatures near −5°C, with the silver iodide becoming more effective as the seed line rises to higher levels in the cloud; and (3) the cloud seeding agent can be released outside supercooled cloud, and when the seeding plume subsequently encounters supercooled cloud, nucleation can occur (ASCE, 2004).

5.2.2.2 Ground Application. Most ground generators utilized in the United States to date have relied upon the generation of silver iodide freezing nuclei. Several different techniques have been developed to generate correctly sized silver iodide particles, including electric arc generators, acetone solution generators, and pyrotechnics. Electric arc generators produce silver iodide particles by passing electricity through an electrode of silver in the presence of iodine. However, the most common type of ground generator in use consists of a tank that holds an acetone solution with a given concentration (usually in the range of 1% to 5% by weight) of silver iodide. Other components include a means of pressurizing the solution tank, a nozzle, and a combustion chamber. Frequently, such systems employ a propane tank with a pressure reduction regulator to pressurize the solution tank as well as serving as a combustible material into which the silver iodide-acetone solution is sprayed. Other systems have been developed which utilize nitrogen to pressurize the solution tank that directly burns the silver iodide-acetone solution (Griffith et al., 1995).

Ground-based generating systems have been developed which are operated either manually or by remote control. Manually operated units are often sited at local residences at low elevations upwind of the target area. Local residents are instructed in the operation of these units and are then called from a central location to turn the generators on or off. Figure 5-6 provides an example of a typical installation. Remotely controlled (radio telemetry, etc.) units are often desirable if suitable residences are

FIGURE 5-6. Manually Operated, Ground-Based Silver Iodide Generator (Griffith et al., 1995).

not located upwind of the target area and to facilitate location of units in higher elevations upwind of the target in order to assure the agent reaches elevations cold enough for nucleation to occur. Both acetone burners and pyrotechnic systems have been developed for remote control applications using radio, telephone, and satellite communications systems. Figure 5-7 provides an example of a remotely controlled acetone system.

Pyrotechnics, similar to the end-burning type described for aerial applications, are also used at surface sites. Again, these units dispense silver iodide nuclei. Racks are built to hold a number of pyrotechnics, which

FIGURE 5-7. Remotely Controlled, Ground-Based Silver Iodide Generator (Griffith et al., 1995).

can be ignited via an automated control system to burn units at a predetermined rate. These units can be operated remotely using the same communications systems as those used to operate the remotely controlled acetone generators. Figure 5-8 provides an example of a typical installation.

5.2.2.3 Advantages and Disadvantages of Aerial and Ground Systems. The most critical portion of any cloud seeding project is the proper delivery of cloud seeding material to the appropriate portion of the cloud. Concentrations of the cloud seeding agent must be adequate to modify a sufficient volume of cloud to significantly affect the precipitation process in the desired manner (Griffith et al., 1995). To date this has been, and continues to be, the greatest challenge in developing a reliable and economically meaningful precipitation enhancement technology.

The complexities of wind flows over mountain barriers and within actively growing convective clouds and cloud complexes make delivery of cloud seeding material and the determination of the fallout of cloud seeding effects very difficult. Major research projects conducted during the last 20 years have spent millions of dollars and applied the most sophisticated measuring equipment in order to determine the transport and dispersion of seeding material and to quantify the effects due to cloud seeding. Results have emphasized that delivering seeding material to the appropriate portions of clouds in sufficient quantities is, at best, limited with current seeding methodologies.

FIGURE 5-8. Ground-Based Silver Iodide Pyrotechnic Dispenser (Courtesy of North American Weather Consultants).

As mentioned in Section 4, to adequately treat supercooled regions of clouds requires producing tens to hundreds per liter of additional ice nuclei after the material has had time to diffuse within the cloud. It has been documented that much of the available supercooled liquid water in clouds, especially in wintertime clouds, is between the 0°C and −10°C levels (Reynolds and Dennis, 1986; Reynolds, 1988; Super, 1999). Based on effectiveness levels of various cloud seeding agents in this temperature range, seeding rates can be determined. A part of this calculation requires knowing how much dispersion occurs from either aerial or ground release of cloud seeding agents.

It is very important that those considering implementation of cloud seeding projects for the enhancement of precipitation should understand that the choice of a cloud seeding delivery system (aerial or ground based) and the accurate targeting of the cloud seeding effects are complex issues. In some cases, several seasons of in-situ measurements of cloud types, liquid water concentrations and temperatures, wind flows and transport mechanisms may be needed before a cloud seeding design can be established.

Again, the project design needs to consider the relative advantages and disadvantages of aerial and ground systems and select the systems that are best suited to meeting the goals of a specific project. Sometimes a combination of aerial and ground systems is a reasonable choice in order to gain many of the advantages of both types of systems while offsetting some of the disadvantages of each of the systems used separately. Aerial systems offer advantages in terms of enhanced targeting of the cloud seeding material into specific regions of the storm or cloud systems, the ability to deliver higher dosage rates into given volumes of cloud, and the ability to seed stable atmospheric situations which may not be possible using ground-based systems. Disadvantages include higher costs (than ground generator operations). It is also difficult to maintain an effective amount of cloud seeding material feeding into clouds affecting a target area of perhaps substantial size over long periods of time (i.e., multiple aircraft may be required). In addition, there are potential hazards of flying in icing or extreme turbulence and possible flight restrictions near major airports and within military operations areas (MOAs).

Advantages of ground generator systems include lower cost of operation, and the ability to operate continuously for extended periods. Disadvantages include an inability to operate ground generators successfully during periods of atmospheric stability (if low altitude dispensers are used), thus, a loss of some cloud seeding opportunities. Greater targeting uncertainty exists, since assumptions have to be made regarding the combined horizontal and vertical transport of seeding material prior to nuclei activation, ice crystal growth, and fallout. The high cloud seeding rates possible with aircraft at effective cloud seeding heights (i.e., temperatures lower than about −4°C) are probably not possible using a ground generator system. Maintenance of remotely controlled generators in isolated locations often requires regularly scheduled maintenance trips involving over-snow vehicles or helicopters.

A remotely operated liquid propane dispenser has received several years of field testing and has been shown to be a reliable method of seeding small volumes of supercooled clouds, even at temperatures near 0°C (Reynolds, 1991). Figure 5-9 provides a photograph of a propane dispenser. However, these units must be situated at elevations known to be in cloud and at temperatures lower than 0°C during winter storms. This might require close proximity to the target area and subjects the seeding effects to the complexities of flow over mountains, including rapid updrafts and downdrafts. Site selection is critical when positioning these dispensers. A further caution in the use of this technique is that no experimental projects conducted to date have demonstrated that precipitation can be increased over a fixed target area using this technique. This technique has, however, been used successfully in supercooled fog clearing operations.

The efficacy of ground-based generating systems in summertime applications is uncertain. Some utilization may be possible in mountainous areas with seedable cumuli. Applications in flat terrain, such as the Great Plains, face complex targeting and cloud seeding rate problems that

FIGURE 5-9. Propane Dispenser (Griffith et al., 1995).

render aerial seeding a much more certain seeding approach in these situations.

5.2.3 Deployment of Cloud Seeding Systems

A project design should consider the deployment of a seeding system. Choices of types of generators to be used as well as seeding rates influence the deployment strategy. Spacing of ground generators or the type of aircraft flight plans to be flown are also an important aspect in specifying a cloud seeding mode. The goal in static mode seeding operations is to achieve concentrations of 1 to 10 nuclei per liter of supercooled cloud. For dynamic seeding, 50–100 or more nuclei per liter may be needed.

5.2.3.1 Dispersion of Cloud Seeding Materials in Winter and Summer Clouds. There has been a concentrated effort during the last 10 to 15 years to document both the horizontal and vertical dispersion of aerial and ground release seeding agents (Super, 1991, 1999). These results indicate that the dispersion rates for AgI from either wing-tip generators or end burning flares on aircraft are about 1 m/s horizontally and 0.1 m/s vertically. Figure 5-10 is a graphical depiction of three seed lines released at −6°C and moving toward a precipitation gauge network. Wind speeds of 50 knots are not atypical for this elevation within wintertime clouds. Note the limited volume of cloud that is actually being affected by cloud seeding. Results would indicate that it is difficult to effectively seed the upwind cloud over a watershed with a single aircraft. Two or more aircraft flying simultaneously may be required.

Ground seeding plumes also have limited horizontal and vertical dispersion. Studies (Griffith et al., 1992) indicate that plumes typically spread out horizontally in a 15- to 20-degree arc, and diffuse vertically to about 1,000 meters above the release elevation. Careful placement and activation of the ground-based generators is required if the AgI nuclei plumes are to reach the −5°C to −6°C level some 30 minutes upwind of the seeding target area. The plumes should disperse horizontally and merge before reaching the target area, yet maintain a minimum concentration of some 10 nuclei per liter.

There has been some research conducted on the dispersion of seeding materials in summer clouds. An inert atmospheric tracer (sulfur hexafluoride, SF_6) was used in experiments conducted in North Dakota summer clouds. The SF_6 was released either at the base or at mid-levels of growing towering cumulus clouds. The dispersion of the tracer was studied through repeated penetrations of the cloud at different altitudes. A real-time SF_6 analyzer on-board the research aircraft documented the presence and concentrations of the tracer gas. Tracer plumes detected at mid-levels of the cloud were found to be relatively narrow and embedded within

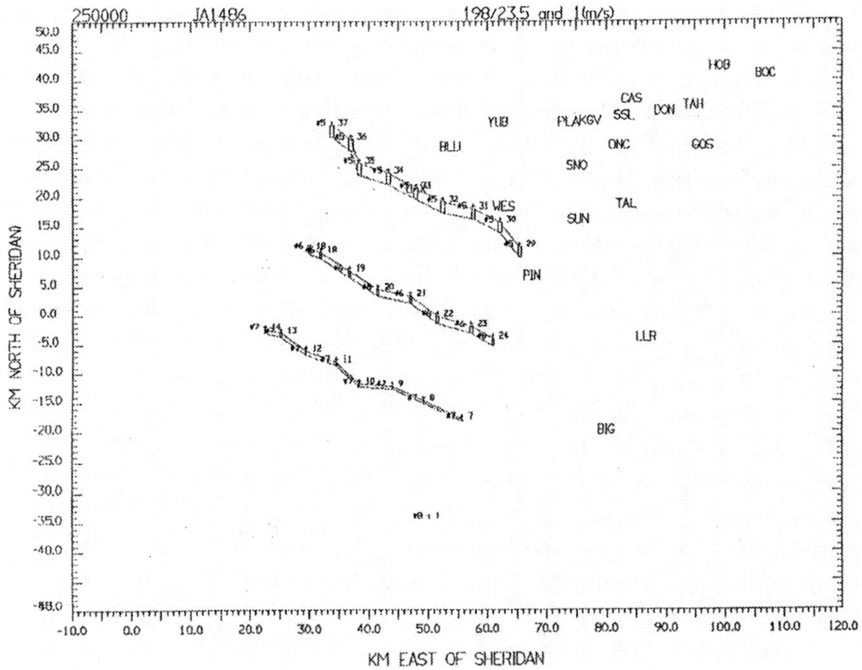

FIGURE 5-10. *Graphical Depictions of Three Seed Lines Produced from Silver Iodide Seeding (Griffith et al., 1995).*

updrafts or downdrafts. Tracer plumes with diameters comparable to the cloud diameters were found in the upper 20% of the clouds. These observations suggested only limited dispersion of the plumes in the clouds, with greater mixing occurring at cloud top (Stith et al., 1986).

5.2.3.2 Aerial Cloud Seeding Modes. Various aerial cloud seeding modes are possible, depending upon the seeding concept, the type of clouds to be seeded, the seeding agent, and the type of dispenser (e.g., acetone-silver iodide burners). Choices of flight levels include seeding at cloud base, in-cloud seeding, or cloud top seeding (Griffith et al., 1995).

Cloud base seeding is frequently used in summertime situations when either silver iodide-acetone generators or burn-in-place pyrotechnics are used. Cloud seeding rates vary in the range of tens to hundreds of grams of AgI per hour. Seeding is usually carried out in the inflow or updraft regions since these regions normally contain the highest concentrations of water vapor and will carry the agent up to the desired range of temperatures in the clouds.

In-cloud seeding is a frequent seeding mode in wintertime projects where the aircraft often fly at or near the melting level (to avoid icing) or in the −5°C to −10°C region where cloud seeding materials begin to operate effectively and where there are often higher concentrations of supercooled liquid water. In-cloud seeding is also utilized in summer operations, although weather radar is usually required (either onboard the aircraft and/or at a ground support site) to permit avoidance of potentially dangerous thunderstorms. A variety of seeding techniques (i.e., dry ice, droppable AgI flares, end-burning flares, and acetone-AgI generators) can be utilized for in-cloud seeding. Cloud seeding rates are normally on the order of 250 to 3,000 grams per km of flight path in cloud when using dry ice or in the tens to hundreds of grams of AgI per km.

Cloud top seeding is employed in both winter and summer operations. Typical seeding targets consist of growing cumulus clouds. Cloud seeding aircraft can penetrate the tops of such clouds in order to locate the updraft regions and subsequently seed these regions (since they frequently contain high supercooled liquid water concentrations). Another approach involves flying just above the cloud top and dropping either dry ice or droppable AgI pyrotechnics into growing tops. Cloud seeding rates are similar to those utilized for in-cloud seeding unless dynamic cloud seeding is called for in the project design. The AgI consumption can range from hundreds to thousands of grams per hour during a dynamic cloud seeding operation. Such situations require close coordination with the FAA. Oftentimes, routine flight patterns to be used on a project can be approved by the controlling FAA Center prior to operations, allowing quick clearances during seeding situations (Griffith et al., 1995).

5.2.3.3 Ground-Based Cloud Seeding Modes. Ground generators have frequently been utilized in wintertime cloud seeding projects or programs in the western United States and other mountainous regions of the world over the past 40 to 50 years (Griffith et al., 1995). Generally, a network of generators is established upwind of a mountain barrier. The number of generators, the spacing between, and their distance upwind of the barrier are determined from considerations of the target area size, expected temperatures in cloud, and the anticipated transport and dispersion of the cloud seeding material. Using ground-based AgI generators requires placing a reliance on vertical dispersion of seeding material caused by atmospheric instability and turbulence associated with naturally occurring storm systems and the barrier. Cloud seeding material is not "shot" or projected into storm clouds, contrary to a frequent public perception of how ground generators function. Output rates are frequently lower than aerial applications, running on the order of 5 to 35 grams of AgI consumed per hour of operation. Some use has also been made of ground-based burn-in-place AgI pyrotechnics to achieve higher output rates from the ground. The latter seeding technique formed the basis of the Santa Barbara II Phase I research project in California (Brown et al., 1974), where 400 grams of silver iodide was consumed every 15 minutes during the passage of a convective band over the seeding site. The cloud seeding in this project was directed at a "dynamic" response, thus, the high seeding rates.

Ground-based rockets and artillery shells loaded with silver iodide or some other seeding agent have been used extensively in several of the former Soviet Bloc countries and China on hail suppression projects. The projectiles are launched with directions from radar and targeted for the supercooled tops of the growing cloud elements. While these methods appear to offer the advantages of both ground and airborne delivery systems in some countries, they are costly and unacceptable for use in regions where there are numerous private or commercial aircraft operations.

5.2.3.4 Possible Studies Related to Proper Targeting of Seeding Materials. If financial constraints allow and program goals indicate a need to do so, targeting during initial seasons of operational programs can be verified by releasing various tracers such as sulfur hexafluoride (SF_6) or indium sesquioxide (In_2O_3) along with the seeding agent(s), or by tracing the ice nuclei themselves with an ice nuclei counter. This has been done in the context of a number of operational programs in the American west, e.g., California (Stone and Marler, 1993; Chai et al., 1993), North Dakota (Boe et al., 1992), and Utah (Super and Huggins, 1992). Such physical verification of targeting lends credence to the methodology employed by the program, and may contribute significantly to the program's longevity. Perhaps more important, improper targeting can be corrected, and program efficacy improved.

A useful, but expensive means of verifying the targeting is to test the snowpack within and outside the target area for traces of those chemicals used in the seeding. Typically, testing is done for silver, as silver iodide is usually a primary component in the ice nucleant used. Work with indium sesquioxide in the Lake Almanor watershed of the Sierra Nevada showed annual silver/indium ratios as high as 8.4:1 at the higher, colder elevations of the target area under westerly flow (Stone and Marler, 1993; Chai et al., 1993). This research suggests that an active ice nucleation process produced new ice crystals, and additional precipitation was linked with more than 85% of the silver detected in the target area snowpack. It was also found that the amount of indium was related linearly with precipitation amount, implying that the indium aerosol particles were being removed directly by the number of cloud droplets the plumes encountered. The indium particle size distribution of the indium-containing aerosols emitted by the source was known, allowing a direct estimate of the amount of precipitation that would have fallen naturally (because a new ice crystal nucleated by the seeding agent has an extremely small chance of collecting both a silver and indium particle). The difference between the total precipitation and the non-seeded component provides an estimate of the amount of seeded precipitation. Case studies conducted in the Nevada/NOAA federal state cooperative program at Lake Tahoe, Nevada, showed that the seeded precipitation can be a large fraction of some precipitation samples. Specialized source-receptor experiments showed that the Ag:In ratio could exceed 10:1, and were correlated with enhanced precipitation rates of up to 6.0 mm h^{-1}.

The presence of silver in snow does not by itself directly translate into quantitative estimates of increases in precipitation. Tracers like indium sesquioxide (In_2O_3) are also seeing increasing use, for the indium is not a nucleant, and when released simultaneously from the same location with the silver-iodide-based seeding agent at known rates, the final ratio of indium to silver in the snow indicates the fraction of the silver that was active as a nucleant, confirming a physical linkage between seeding and ice production. This kind of physical process (not statistical) evidence strengthens the argument that seeding was the cause of the increased precipitation.

Another tool that is now available is new, efficient, cesium-tagged ice-nucleating seeding agents. Although the concept is not new, the efficiency of the seeding agents now makes them a viable option for use in operations. The additional expense of adding cesium to the seeding agent is not great, especially given the added confidence it provides in terms of targeting verification. The expense, however, of silver and cesium and silver analyses can be significant. The newer formulations have been shown to outperform older "standard" AgI seeding agent yields by as much as a factor of 30 at temperatures of −8°C. Applications are numerous, including: testing of new generator locations or techniques while minimizing impacts on existing operational seeding programs; detecting impacts on control areas and extended areas downwind of primary target areas; measuring the (contamination) impacts of neighboring seeding programs; and determining the relative contributions of aircraft versus ground-based seeding components of individual programs. Each of these applications involves detecting cesium to uniquely identify the source of the seeding materials, and unlike the indium control tracer, which requires the duplication of seeding agent release facilities, cesium can be added to seeding materials and released without the need for additional operational apparatus. Cesium has a natural background level equal to or lower than silver; so detection of cesium and silver in samples located downwind of their release point provides physical evidence of the source of the materials. Cesium can also be readily incorporated into modern pyrotechnics.

5.3 INSTRUMENTATION

There are two basic needs for meteorological or hydro-meteorological information in supporting a cloud seeding project. Real-time data are needed to support decision making and monitoring functions. Past cloud seeding research has identified certain situations during wintertime or summertime precipitation episodes that respond favorably to cloud seeding, while others appear to either not respond or to respond unfavorably. Consequently, the project design must consider the development of "seeding criteria" for use on the project in order to conduct seeding only during those periods when a favorable response is anticipated (ASCE, 2004).

Supercooled liquid water is a necessary but not altogether sufficient condition for initiating a cloud seeding project. Research has shown that liquid water is highly variable both in space and time in wintertime storms and cannot be correlated well with any specific meteorological variable (Reynolds, 1988). This is also true of summer convective clouds where entrainment of dry air can rapidly erode a convective element's liquid water before cloud seeding can utilize this water. In-situ or remote measurements of supercooled liquid water are therefore desirable to allow informed decisions to be made.

As discussed in the previous sections, delivering cloud seeding material to the appropriate place and time in a cloud is very difficult. Information on atmospheric stability and vertical wind profiles is required to assess the potential of ground released seeding material to reach appropriate levels in cloud and to determine the expected fallout region of enhanced precipitation (i.e., targeting).

Project monitoring also includes a need for information concerning possible hazardous situations during which cloud seeding should not be performed. These situations should be established in project suspension criteria.

Information on movement of storms, the likelihood of a storm affecting the project target area, the likely ending time of precipitation in the target area (i.e., forecasting), etc., is also an important component of real-time decision making. This type of information provides an important input to project planning in establishing personnel schedules, maintenance schedules, etc.

Post-project assessments of results represent a second basic need for data from various types of instrumentation. It is quite important to the continued viability of an operational cloud seeding project to consider methods of assessing and reporting the effects of the seeding. Many projects without some method of assessment or indication of the achievement of a positive effect are abandoned within three or four years due to lack of support.

An assessment of the effects of cloud seeding in an operational project where every favorable seeding event is seeded is not a simple matter. Ideally, a project design completed prior to the initiation of the project will consider, among other topics, the question of assessing the effects of cloud seeding. Types of instrumentation of potential value in satisfying each of these general requirements (i.e., real-time and post-project) are examined separately in the following sections.

5.3.1 Real-Time Decision Making and Monitoring Instrumentation

There exists an array of hydro-meteorological instrumentation of potential value in the day-to-day conduct of a cloud seeding project. A considerable amount of information is in place at many locations serving other functions, and it can be utilized in cloud seeding projects. Examples include the National Weather Service (NWS) network of data collection, data assimilation, data processing, and data dissemination functions (Griffith et al., 1995). Depending upon the location of the target area in relation to existing data collection points and the specific needs of a particular project for certain types of instrumentation (as established in the project design), it is likely that some additional project-specific instrumentation will need to be acquired, installed, and operated in support of the project. Such installations may be of use in terms of real-time project operations as well as post-project assessments. The various types of instrumentation of potential value to a cloud seeding project for real-time decision making and monitoring functions are covered below.

5.3.1.1 Available National Weather Service Data. The NWS collects a variety of meteorological data and issues weather forecasts to a diverse group of users. Basic data as well as analyzed data, projections, and forecasts are available to cloud seeding projects from the NWS or private companies that provide data via the Internet. Types of data available by these means include rawinsonde data from a national network of stations on a twice daily basis. As mentioned previously, such data, if collected in very close proximity to the project area, are useful in trying to establish the "seedability" of a given situation as well as "targeting" the effects of cloud seeding. Hourly observations from a large number of reporting stations are also available. These observations include information on sky cover, surface winds, temperature, humidity, and present weather. This type of information is also quite useful in developing a knowledge of the structure of a particular storm as it affects a given target area.

Analyzed data include surface pressure patterns and constant pressure charts at upper levels, as well as forecasts of these pressure fields at six- or 12-hour intervals out through 72 hours and beyond. Current weather can best be observed via weather radar and satellite images. The NWS or the same companies that provide weather data can provide images from both local weather radar and geostationary satellite. Such images can be stored and displayed in sequential fashion on video monitors allowing extrapolation of weather features into the target area ("Nowcasting").

Project suspension criteria may require the monitoring of special meteorological or hydrological data. These might include telemetered river and reservoir levels, and precipitation data from special telemetered gauges designed to monitor heavy precipitation and to automatically alert water managers and the NWS of threatening weather. The NWS also issues Special Weather Statements and weather Watches and Warnings as conditions warrant. These can be obtained via NOAA Weather Radio, the NWS, or through the same data stream providing weather data. For winter projects it may be necessary to contact the U.S. Forest Service via a special recorded phone message that warns of any avalanche hazards.

The NWS has replaced its conventional weather radars with Next Generation Weather Radars (NEXRAD) (Baer, 1991). This replacement process was completed in 1996, with approximately 136 NEXRAD sites in the contiguous United States (Klazura, 1993). The NEXRAD radar (WSR-88D) is a Doppler radar, which provides high-resolution reflectivity and velocity information. The NWS has developed approximately 39 categories of analysis products. A number of these products are available in near real-time from a variety of Internet providers, including data available directly from the NWS. Success of the Collaborative Radar Acquisition Field Test (CRAFT) project, led by K. Droegemeir of the University of Oklahoma, should enable users to access the full-resolution data base in near real time from any NEXRAD site via data compression techniques. Volume scans from TITAN-equipped project-dedicated radar (TITAN is discussed in section 5.3.1.3) and NEXRAD radars take approximately the same amount of time (i.e., about 5 minutes); therefore, if data are acquired from NEXRAD radar shortly following the completion of each volume scan they can be used as effectively as TITAN-equipped project radar to make real-time seeding decisions.

The NWS has utilized weather satellites since the 1960s to provide information on cloud cover. Weather satellites have become increasingly sophisticated with time. Both orbiting and geostationary satellites (satellites that can remain in a "stationary position" over the earth's equator) are used. Different sensors are routinely carried on these satellites providing, for example: visible images, infrared images, and information on atmospheric moisture. Data are available from the geostationary satellites approximately every 10 to 15 minutes. Satellite information is useful in the conduct of cloud seeding projects to determine the location and movement of the storm or cloud systems of interest. The infrared data capability provides this information during nighttime hours and also can be used to estimate cloud top temperatures, which are of interest in assessing the potential seedability of winter storms. The data provided by satellites can also be analyzed to provide estimates of different parameters such as the water content of the atmosphere (Guillory et al., 1993), which can be of interest in the real-time conduct of cloud seeding projects.

An evolving technique uses multi-spectral analyses of satellite imagery. With timely satellite imagery, it is possible to calculate the evolution of the effective radii of convective cloud particles with respect to temperature, which provides information about precipitation-forming processes in the clouds (Rosenfeld and Lensky, 1998). The usefulness of polar orbital satellites is limited by the number of times one passes over the area of interest. Usefulness of both polar orbital and geostationary satellites is limited by the horizontal resolution of the sensors.

5.3.1.2 Special Project Precipitation Gauges. Quite often, the number and/or location of existing NWS and other precipitation gauges are inadequate in terms of a specific cloud seeding project's requirements. Typical uses of special precipitation gauges are: determining the onset of precipitation and thus possible cloud seeding potential, monitoring for excessive precipitation periods, and collecting precipitation data for post analysis. Types of gauges available for installation include weighing type precipitation gauges and tipping bucket gauges. The type preferred mainly depends on whether it is a summer or winter project and the intensities of precipitation that can be expected in a given target area. Sometimes it is desirable to provide data in real time to the project meteorologist, in which case telemetry of some type is needed.

Special provisions are required if the gauges are to be used to measure snow. This includes mounting on towers, shielding the orifice from wind to increase catch efficiency, and making a provision to melt the snow. Weighing gauges are the only satisfactory method of measuring the water equivalent of snowfall. Tipping bucket gauges must be heated to operate in snow, but it is well known that the heat rising from the gauge is sufficient to raise the trajectory of ice crystals, causing them to pass over rather than into the gauge and thus under-sampling precipitation. An antifreeze solution must be added to the weighing gauge to melt the snow as it is collected. An evaporation suppressant (usually mineral oil) must also be placed into the antifreeze solution to prevent the antifreeze from evaporating. Special mixtures of methanol and glycerin, which are non-toxic, have been tested and found to work as well as glycol (toxic) mixtures (Price and Rilling, 1987).

A second problem in snow country is capping of the gauge orifice by accumulating snow. It is recommended that the gauge have at least a 30-cm orifice and that the sidewalls of the gauge be minimized to avoid snow sticking to the side of the gauge and bridging over the orifice (Griffith et al., 1995). A newer design, the optical rain gauge, offers an attractive alternative in measuring snow that may have some advantages over the weighing or tipping bucket gauges (DeFelice, 1998).

Gauge resolution is also important. For wintertime projects, the gauge should be able to resolve precipitation to within 0.3 mm. This is on the order of what seeding produces in about an hour. Because of the requirement for high resolution and a large orifice, servicing of the gauge will be frequent unless the catch can be automatically drained and a fresh antifreeze charge placed into the gauge. Figure 5-11 provides a photograph of a weighing type rain/snow gauge, which automatically drains and recharges itself, requiring no visits during a winter season.

FIGURE 5-11. Weighing Bucket Recording Precipitation Gauge (Courtesy of North American Weather Consultants).

5.3.1.3 Special Project Weather Radar. Weather radar is available in a variety of wave lengths (K-, X-, C-, S-band) each of which can serve a slightly different function in cloud seeding projects. K-band (1 cm) radars, pointed vertically, can provide information on cloud top heights in winter storms. They may also be sensitive enough to directly observe cloud seeding effects. Radars have been developed that are Dopplerized (a technique which provides wind direction and speed information in addition to the precipitation intensity), providing both reflectivity and radial velocity (Pasqualucci et al., 1983). K- and X-band (3 cm) radars can detect light to heavy snowfall, and light to moderate rain, but are attenuated during heavy rainfall. Dual-polarization radars can be used to track chaff (a tracer) to follow dispersion of seeding material. By far the most commonly used radars in cloud seeding projects are C-band (5 cm) radars. They provide sufficient sensitivity in rain events but decreased sensitivity in snow events. They also can be Dopplerized. S-band (10 cm) radars are superior in heavy rain and hail situations but may lack sensitivity in measuring snow.

Radars are normally operated in either PPI (plan position indicator) or RHI (range height indicator) modes. The PPI mode provides a horizontal depiction of the precipitation, which the radar sees out to a range of perhaps 180 to 360 km. Multiple PPI scans at different elevation angles can be accumulated over a 4- to 5-minute period forming a volume scan. An RHI mode provides a vertical presentation of precipitation the radar sees along a certain azimuth from the radar. Radar digitizing is available and can be specified for a project if storage and retrieval of data are of benefit. Project-specific weather radars are frequently used in summertime cloud seeding projects. They are used somewhat less frequently in wintertime projects, especially if aircraft are not used. In summertime operations, they can be used to keep track of high reflectivity, potentially hazardous areas (i.e., convective bands or probable hail regions) to be avoided by the cloud seeding aircraft, as well as identifying areas of new echo development of potential interest as cloud seeding targets.

The utility of weather radars for use in aerial seeding operations can be enhanced considerably if they are equipped with an aircraft tracking capability, which utilizes a radio modem to transmit a Global Positioning System (GPS) aircraft location to a ground-receiving site (typically the project radar). When the project radar is equipped with such a unit, the operator can visually keep track of the location of the cloud seeding aircraft in relation to weather echoes as well as ground terrain (often of special interest in mountainous areas). In some cases, radar information can be utilized as input to suspension criteria considerations. Figure 5-12 contains a photograph of a typical field radar installation.

An example of useful automated radar analysis is the family of capabilities for the automatic identification and tracking of individual storms in systematically recorded volumetric radar data. An example of this is the package called TITAN (Thunderstorm Identification, Tracking,

FIGURE 5-12. Self-Contained, Project Dedicated Weather Radar Installation in the Field (Courtesy of Weather Modification, Inc.).

Analysis, and Nowcasting) developed in South Africa; versions have been used and made available to the public by the National Center for Atmospheric Research (NCAR) Research Applications Project (RAP). TITAN is described in Dixon and Wiener (1993). Storms are identified on each volume scan (a collection of individual PPI scans taken at different elevation angles in a short period of time) as volumes enclosed by an envelope composed of a specified surface of threshold reflectivity, and a complex algorithm associates a storm cell on one scan with its position on the next. For each volume scan, parameters, such as storm height, volume, mass, etc., are estimated for each storm, and the time history of these constitutes a description of their life cycle. Storm outlines can be overlaid on the radar display at each scan time, for a number of past scans in one color, for the present in another color, and for a number of extrapolated future times in yet another. The prognostic cell positions are computed from a forecast algorithm. The TITAN software can also be used to display the location of the seeding aircraft in relation to the weather echoes.

The National Severe Storms Laboratory has developed a new, improved NEXRAD cell-tracking algorithm (known as SCIT) that will be implemented on the NWS Advanced Weather Interactive Processing System (AWIPS) computer workstations.

5.3.1.4 Special Project Rawinsondes. Rawinsondes (weather balloon-borne instrument packages) are a common addition to an operational cloud seeding project. As previously mentioned, balloon-borne (helium- or hydrogen-filled) instrument packages transmit back (via radio) pressure, temperature, and humidity data to a surface site. By tracking the balloon, using either triangulation with LOng RAnge Navigation (LORAN), GPS, or radio signal strength (radiotheodolite), one can obtain a vertical profile of both wind direction and speed.

Data are required from the surface to varied heights (up to 15 km), depending upon the project requirements. Observation times are also on an as-required basis, varying from 3-, 6-, 12-, or 24-hour releases during active weather conditions. The location of a rawinsonde field site is not as critical for summertime projects as for wintertime projects. For winter projects the site should be located upwind of the target area especially in mountainous regions. In fact, for large mountain barriers, two rawinsonde sites may be needed; one on the upwind edge and one near the crest of the barrier to adequately measure the complexities of the flow. Figure 5-13 provides a photograph of a typical rawinsonde facility. Near real-time rawinsonde data are valuable input in a variety of ways.

In summertime, the convective potential of the air mass can be determined. This same sounding can be used as input to numerical cloud models, which can provide information on the dynamic cloud seeding potential for the particular air mass. Data from rawinsondes can be utilized to initiate suspension if, for instance, extreme instability is indicated which would suggest high likelihood of hail formation.

In wintertime, the rawinsonde data can also indicate whether any low-level stability exists in the atmosphere, which may limit or preclude ground-

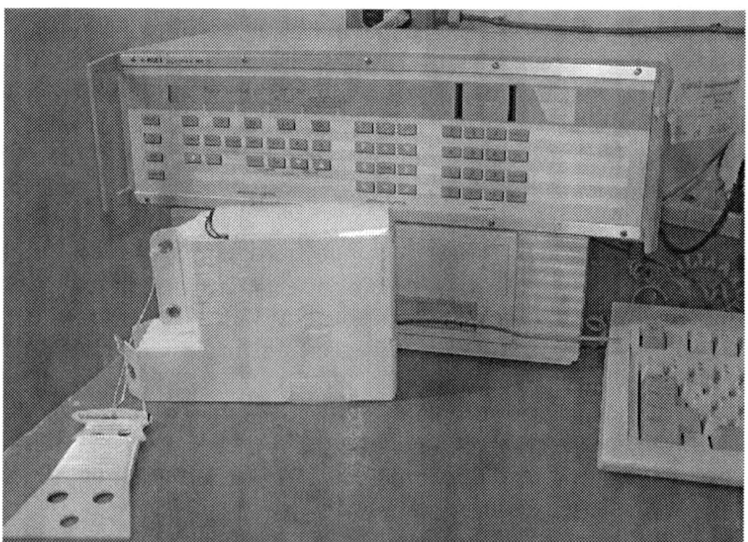

FIGURE 5-13. Rawinsonde Receiver System (Courtesy of Weather Modification, Inc.).

based seeding with AgI. Targeting guidance is also available from upper-level wind data. Data from upwind and, if available, crest or downwind sounding data can be input to simple kinematic and microphysical models to predict the transport, nucleation, and fallout of seeded ice crystals.

5.3.1.5 Supercooled Liquid Water Observations. One of the observations particularly useful in the conduct of wintertime snowpack enhancement projects is the observation of supercooled liquid water (SLW). It is also an important parameter in summer rainfall enhancement projects but can only be derived from in-situ aircraft measurements. This will be discussed in the next section.

An important instrument applied to the measurement of supercooled liquid water in winter mountain clouds is the dual channel microwave radiometer (Hogg et al., 1983a). This research tool provides valuable information on both the SLW and water vapor passing over a given mountain range. The instrument passively detects the presence of both condensed cloud water and water vapor in a narrow beam above the radiometer. When operated at elevations where the temperature is lower than 0°C, only SLW due to cloud droplets is observed. When operated at lower levels with higher temperatures, melting crystals and cloud liquid at temperatures above 0°C contaminate the system's data. This instrument cannot determine the location of the liquid water in the vertical. However, another instrument using a laser can help locate where in the vertical the liquid water may be. A lidar (Sassen, 1992) has been shown to be useful in this regard if precipitation is not intense. Figure 5-14 provides an example of a radiometer and lidar installed side by side on a research project.

FIGURE 5-14. Microwave Radiometer, left, and Lidar, center (Griffith et al., 1995).

Another remote-sensing method (based upon satellite information) has been developed to help identify the presence of supercooled liquid in cloud tops (and thus the potential for glaciogenic seeding and identifying seeding signatures). This method has been described by Woodley et al. (2000).

A more practical and inexpensive approach to measuring SLW is the installation of an icing rate meter at mountaintop levels. The device is similar to ice detectors used on aircraft or by telephone companies operating equipment subjected to severe rime icing. The detector has a small 25-cm probe protruding from a small half hemisphere. The probe vibrates at a known frequency. When ice accumulates on the probe tip (i.e., a process known as accretion), it changes the vibration frequency, causing a heater to switch on at a predetermined ice mass, melting the accumulated ice. By knowing the number of deicing cycles, the wind speed, and the mass of ice required to cycle the heater, one can calculate the liquid water content (Hindman, 1986). This type of detector has been used successfully as a real-time indicator of SLW. Mounted on a mountaintop having commercial or other power source, the data can be telemetered via satellite or telephone to the project operations center. It is useful to have additional observations at this same site. These would include temperature, wind speed, and direction. Temperature would be useful to determine a particular cloud seeding agent's activity level, and wind speed to quantify liquid water amounts (Griffith et al., 1995).

5.3.1.6 Special Project Cloud Physics Instrumentation. Specially instrumented cloud physics aircraft are frequently used in research-oriented cloud seeding studies, but their application to operational cloud seeding projects has been rather limited. The primary consideration has been one of cost, since a fully equipped aircraft can be very expensive to configure and to operate. One such specialized aircraft is the University of Wyoming's King Air cloud physics aircraft shown in Figure 5-15. The King Air was utilized on the Bureau of Reclamation's Sierra Cooperative Pilot Project (SCPP), the High Plains Cooperative Program (HIPLEX), and the North Dakota Thunderstorm Project (NDTP). The significant costs are justifiable on research projects where the need to understand basic processes at work in cloud systems of a given area is of utmost importance. A compromise situation is possible, however, especially when a cloud seeding project utilizes one or more cloud seeding aircraft. This compromise consists of the installation of relatively basic cloud physics instrumentation on the seeding aircraft. The parameter of perhaps greatest interest is supercooled liquid water, which is directly related to the seedability of a cloud system. Other parameters of interest include some measure of the concentration of ice particles (an indication of how naturally efficient the system is), temperature, and some water vapor measurement (i.e., dew point or relative humidity).

FIGURE 5-15. University of Wyoming King Air Cloud Physics Aircraft (Griffith et al., 1995).

5.3.1.7 Other Instrumentation and Equipment. As has been repeated throughout this section, it is critical that confidence be gained that the cloud seeding agent being used is making it to the appropriate regions of the intended cloud in sufficient concentrations to be meaningful. Obviously this applies more to ground-based seeding than to aerial seeding. Several methods are available to do this. Tracer studies during the first one to two years of a project may be invaluable in building confidence that seeding material reaches the appropriate SLW-rich cloud levels the majority of the time. Tracer studies require both a good understanding and observation of the general wind flow and the release of an aerosol that can simulate the trajectory of the seeding agent.

It is critical that the environmental winds be observed during these experiments. Rawinsondes have already been discussed. More continuous observations can be obtained from sodars and wind profilers (Weber and Wurtz, 1990). Sodars, or acoustic sounders, are much less expensive but provide vertical profiles of the horizontal winds within only the lowest kilometer above the device. Profilers (Hogg et al., 1983b) provide winds to the top of the troposphere but at a somewhat degraded resolution in the lowest levels compared to the sodar. Doppler weather radars and properly equipped research aircraft can also provide horizontal wind measurements but their applications are somewhat limited over mountainous terrain due to terrain blocking and aircraft flight level restrictions, respectively (Griffith et al., 1995).

There are at least three methods currently available for tracing the trajectory of aerosols to simulate cloud seeding. The first is the use of an ice-nucleus counter (Langer, 1973) to directly measure the presence of AgI. This detector can be mounted either in a van or on an aircraft. It basically acts as a portable cloud chamber cooled to −20°C. When ice nuclei are injected into the chamber, crystals grow and fall out onto an acoustic detector. By counting the number of acoustic emissions one can obtain a qualitative estimate of ice nuclei concentrations. The response time of this system is slow, thus smearing the plume. It can also be quite difficult to operate unless specialized training is obtained. Its main advantage is that it is sampling the actual seeding plume and not a surrogate (Griffith et al., 1995).

The second tracer method is through release and detection of rare gases that can be measured down to parts per trillion (PPT). One such tracer gas used extensively in air pollution work is sulfur hexafluoride (SF_6). Fast response analyzers (Benner and Lamb, 1985) provide a means to measure the gas concentrations in PPT. It is applicable for both winter and summer cloud seeding projects and has been used successfully on both (Stith et al., 1986; Griffith et al., 1990, 1992).

A research technique has evolved using circularly polarized X-band radar (Martner and Kropfli, 1989). Hydrometeors scatter radiation both horizontally and vertically, and this type of radar measures both components of the scattering. By releasing small pieces of reflective material (chaff) it is possible to track the material through a cloud, even if the cloud is precipitating moderately. This is because the depolarization signal from the chaff is much greater than that from precipitation particles. By scanning through the cloud in which chaff has been released, it is possible to get a three-dimensional view of transport processes in cloud. The chaff does not remain suspended in the cloud but has a fall speed of about 30 centimeters per second. This gradually separates seeding materials from the chaff. Experimentation continues using other material that would not have such high terminal velocities.

5.3.2 Measurements of Potential Value in Post-project Assessments

The other major requirement for instrumentation measurements other than real-time decision and monitoring functions is data for post-project assessments of the cloud seeding effect. There are again a variety of measurements that have been investigated and used for this purpose (Griffith et al., 1995).

Assessment of the effects of seeding in operational cloud seeding projects presents a challenge to man's ingenuity. The question of how much rain or snow would have occurred in a given situation without seeding is a straightforward, yet deceptively complex issue. The culprit is the normally large natural variability in precipitation from day to day, month to month, and year to year. Imposed on this large natural seasonal variability may be a seeding signal of 5% to 10%, which proves to be within the noise level in many cases. An approach utilized in research projects involves randomization, whereby normally one-half the seedable events are left unseeded for comparison with the seeded ones. Statistical tests can be applied and if the seeding signal is large enough and the experiment is conducted long enough, some statements can be made concerning cloud seeding success. Sponsors of operational projects are normally unwilling to forgo one-half the potential benefit of a cloud seeding project. Consequently, other assessment approaches are required. The most common approach is a target and control comparison (Dennis, 1980) where a control area is selected such that the effects of seeding should not affect it (this selection is ideally made prior to the beginning of seeding). A historical period is selected during which a regression analysis can relate measurements such as precipitation, snowpack water content, or streamflow to similar measurements in the target area. This historical period should not include any periods with previous seeding in either the control or target areas. These regression relationships are then used to predict the amount of natural precipitation, etc., in the target area during seeding (from the control area measurements) for comparison to the observed precipitation in the target area. Griffith et al. (1990) provides an example of the application of this assessment technique to a winter orographic cloud seeding program being conducted in Utah.

Unfortunately, target-control analyses can suffer from changes in relationships from the target to control from year to year and especially for various prevailing wind (storm track) regimes. These types of assessments are more prone to statistical errors than are purely randomized experiments. Historical comparisons via simple regressions are intended to reveal systematic changes in target-control relationships, and might mix effects of cloud seeding with climatic fluctuations and changes due to other causes. Use of additional controls, at different geographic directions and orographic situations may reduce this risk (e.g., Ben-Zvi and Langerman, 1989).

Other assessment techniques have been utilized to determine possible cloud seeding success. These techniques are generally concerned with documenting the physical links in the chain of effects from cloud seeding to increased precipitation at the ground. Even if only limited physical studies such as transport and diffusion are performed, they may provide more credibility to the statistical results. Certain effects due to cloud seeding are hypothesized (such as a decrease in supercooled liquid water and an increase in ice crystal concentrations) and these hypotheses are checked via physical measurement. The various types of measurements of potential value in seeding assessments are discussed in the following sections (Griffith et al., 1995).

5.3.2.1 Precipitation Gauge Data. Measurements from precipitation gauges provide the most common form of data from which target-control assessments are made. It is highly desirable to establish target-control relationships to be used in an evaluation prior to the beginning of the seeding activities to avoid possible bias in the selection of control areas after the fact. Typically, monthly or seasonal regressions are developed in such assessments. A common problem encountered in performing such assessments in mountainous areas of the United States is a lack of gauge sites at higher-elevation locations. Another problem is one of historical gauge movements or changes in the type of gauge, which can alter the precipitation measurement, making long periods of record incompatible. Installation of additional precipitation gauges for the project is only useful if randomization is employed, otherwise there would be no historical database from which regressions could be developed. Information has already been described on the necessity for high gauge resolution, depending on the length of the experimental units outlined in the project design. In parts of the United States, assessments are also hindered by the designation of certain higher-elevation areas as "wilderness areas" meaning accessibility is limited if not excluded altogether. Analysis of ground gauge records need not be limited to rainfall amounts. Enhancement of rainfall intensity, duration, and spatial correlation can help in detecting and assessing seeding effects (e.g., Sharon, 1978; Gagin and Gabriel, 1987; Ben-Zvi, 1988, 1989; Ben-Zvi and Fanar, 1996).

5.3.2.2 Remote Sensor Data. Weather radar measurements (overlays, time-lapse photography, or digitized data) have been utilized in assessments of summertime rainfall enhancement projects. Weather radar has not yet provided quantitative measurement of snowfall (especially in mountainous areas) and therefore has not been used successfully in evaluating the end result of wintertime orographic cloud seeding projects. This is because the radar is much more sensitive to a few large snow particles than to many small snow particles which are commonly produced by cloud seeding. The use of radar data from sites established for the cloud seeding project suffers from the same lack of any historical unseeded database for comparison purposes. To study the behavior of the clouds in two areas, some analysis of unseeded cumulus clouds outside a seeded region may be warranted.

Information on echo size, echo height, height and timing of first echo, echo intensity, and echo duration for the seeded and unseeded clouds may suggest certain significant systematic differences. Digitized radar data can be used to crudely estimate rainfall rates at cloud base. These calculated rates are often a more accurate indicator of rainfall in isolated thundershowers than widely spaced precipitation gauges. Dual-polarization and dual-wavelength radars offer better possibilities for hydrometeor type identification and precipitation (rain and snow) measurement, but are still in a research mode. There has been some improvement of NEXRAD snow accumulation algorithms (Super and Holroyd, 1998). There was also a comparison of NEXRAD rain accumulation algorithms with numerous rain gauge measurements for different types of horizontal gradient events (Klazura et al., 1999).

Data from remote sensors have been used successfully in other projects, mostly for cloud seeding research. Microwave radiometers can continuously detect water vapor and liquid water along a viewing path; these instruments may be used to monitor weather conditions. For example, when placed upwind of a mountain barrier, they can signal the presence of supercooled liquid water and thus an opportunity for orographic cloud seeding for conversion to precipitation. Lidars with polarization capability have been used to detect the location and phase (water or ice) of clouds. Wind-profiling radars and Radio Acoustic Sounding Systems (RASS) have been operated in a demonstration network over the central United States and Alaska by NOAA's Forecast Systems Laboratory; these data are available in near real time via the Internet.

5.3.2.3 Cloud Physics Data. If cloud physics data are available on a project, an analysis of the data can often provide a physical understanding of the sequence of events in the production of precipitation. It may be possible in these analyses to interpret the presence of a possible seeding effect. For instance, a systematic decrease in supercooled liquid water content and a corresponding increase in ice crystal concentrations following cloud seeding would strongly suggest a cloud seeding effect in the cloud, although the resultant impact on the precipitation production at ground level would not be explicitly addressed. In winter projects, ground level observations of ice crystal types (referred to as habits) and the degree of riming (collection and freezing of supercooled water drops upon ice crystals) can provide additional information of the likely regions of ice crystal formation (temperature dependent) and the relative efficiency of the system. These data can then be analyzed before, during, and after seeding to investigate possible differences that may be attributable to cloud seeding (Warburton and DeFelice, 1986 provides an example). The same instruments used on aircraft for measuring ice crystal concentrations (PMS 2D-C probes) can be used at the ground to count and size ice crystals. Either aspirating a fixed-site probe or mounting the probe on a vehicle and driving it through the desired area accomplish this. Microphotographs are needed to assess rime characteristics on crystals (Deshler, 1988).

5.3.2.4 Streamflow Data. Streamflow measurements, typically compiled by the U.S. Geological Survey, can be utilized to make cloud seeding assessments. The Kings River Project in California (Henderson, 1981) has utilized a target/control assessment based on historical seasonal runoff amounts having correlation coefficients in the 0.95 to 0.98 range (Hastay and Gladwell, 1969). Even with high correlation coefficients, tens of years may be required to assess the cloud seeding effect with a statistical significance in the 0.05 range since only one measurement per seeded year is acquired. One problem often faced in streamflow assessments of this type is either the lack of unimpaired runoff measurements or a change from an unimpaired measurement point to one with new dams or diversions constructed upstream at some point in either the historical or the seeded period. Some techniques are available to calculate natural flows taking such factors as evaporation from lake surfaces into account. Another problem is the potential for carry-over flow from one year to another, which impairs the independence between different annual records, thus making the assessment more complicated (e.g., Ben-Zvi and Langerman, 1995).

*5.3.2.5 **Snow Course Data.*** Following Church's (1918) historical work on the development of techniques to measure snow water content in the Reno, Nevada, area, an extensive snow course measurement network has been organized in the Western United States as well as numerous foreign countries. Beginning in December or January and continuing through April or May, monthly measurements of snow depth, water content, and density are acquired at a large number of stations/sites. More recently, the Natural Resources Conservation Service using their SNOTEL system has automated such observations. This automated system now provides several observations per day; both snowpack water content and accumulated precipitation data are available via the Internet. These data can be utilized either separately or combined with precipitation data to perform cloud seeding assessments using the target and control approach. SNOTEL/snow course measurements can fill some of the voids in precipitation gauge measurements in higher mountainous areas mentioned previously. Target and control measurement sites should be at or near the same elevation since melt rates can vary by elevation, which could lead to the development of lower correlations than might otherwise be possible. Movement of snow course measurement sites over the years must be kept in mind. This may have a significant impact on subsequent relations between target and control areas.

5.3.2.6 Snow Sample Data. Occasionally, samples of newly fallen snow are collected for an analysis of silver content. This is an evaluation technique encountered more frequently in research projects due to the expense involved. Snow samples collected prior to cloud seeding or from non-seeded storms are analyzed to establish the natural background silver content (if measurable with available analysis techniques) for comparison with snow samples taken from seeded storms. This technique is only valid for projects using silver iodide as the cloud seeding agent, although some analysis techniques are applicable to other possible cloud seeding agents as well (i.e., lead iodide). Several analysis techniques have been developed for use in such analyses, including neutron activation, proton excitation, and flameless atomic absorption. An example of an analysis of the downwind transport of silver iodide outside of primary target areas is given by Warburton (1974). Warburton et al. (1996) demonstrates how trace chemical assessment techniques strengthen traditional target and control precipitation analyses.

A modification of this trace chemistry assessment technique involves the simultaneous release of a control aerosol along with an active seeding aerosol (Warburton et al., 1996). Such tracers have properties very similar to the seeding agent, with the key exception that it does not nucleate ice. It is insoluble in water, has an extremely low natural background in precipitation, and is only removed from the atmosphere by passive precipitation scavenging mechanisms. Both the seeding agent and tracer are transported and scavenged in very similar manners when conditions are not conducive for effective seeding. Given similar release rates, detecting the same concentrations of silver and indium in precipitation samples at downwind locations indicates that the two aerosols were most likely removed from the atmosphere solely by scavenging. On the other hand, when sufficient SLW exists and temperatures are cold enough for the active seeding material to nucleate new ice crystals, the ratio of silver to tracer in target area precipitation samples can be much greater than unity. This indicates that some fraction of the seeding material was directly responsible for the nucleation of ice crystals that eventually produced additional snowfall.

5.4 STATUS OF PRECIPITATION ENHANCEMENT TECHNOLOGY

The current status of precipitation enhancement technology has been addressed by four of the major organizations that have dealt with weather modification during the past 50 years: ASCE, Weather Modification Association (WMA), American Meteorological Society (AMS), and World Meteorological Organization (WMO). The position statements of these organizations on weather modification are summarized below.

5.4.1 American Society of Civil Engineers

The ASCE Policy Statement #275 was approved by the ASCE Board of Direction in May 2003 (ASCE, 2003).

Policy

The American Society of Civil Engineers (ASCE) supports and encourages the protection and prudent development of the Nation's atmospheric water (also known as "weather modification" or "cloud seeding") for beneficial uses. Sustained support for atmospheric water data collection, research, and operational programs, and the careful evaluations of such efforts including the assessment of extra-area and long-term environmental effects, is essential for prudent development. ASCE recommends that the results and findings of all atmospheric water-management programs and projects be freely disseminated to the professional community, appropriate water managers, and to the public.

Issue

Atmospheric water management capabilities are still developing and represent an evolving technology. Longer-term commitments to atmospheric water resource management research and operational programs are necessary to realize the full potential of this technology.

Rationale

Water resources worldwide are being stressed by the increasing demands placed upon it by competing demands generated by population growth and environmental concerns. As a result, nations have become more sensitive to year-to-year variations in natural precipitation. The careful and well-designed management of atmospheric water offers the potential to significantly augment naturally occurring water resources, while minimizing capital expenditures for construction of new facilities. New tools such as radar and satellite tracking capabilities and other imaging devices, atmospheric tracer techniques, and advanced numerical cloud modeling offer means through which many critical questions might now be answered. Continued development of atmospheric water-management technology is essential. ASCE has developed materials providing guidance in the use of atmospheric water-management technology with weather modification organizations for dissemination to local communities and governments as well as state, regional and international interest.

5.4.2 Weather Modification Association

The WMA's Capability Statement on winter and summer precipitation augmentation is provided below. Also see http://www.weathermodification.org. for more details on the 2005 version of their statement.

Winter Precipitation Augmentation

The capability to increase precipitation from wintertime orographic cloud systems has now been demonstrated successfully in numerous "links in the chain" research experiments. The evolution, growth and fallout of seeding-induced (and enhanced) ice particles have been documented in several mountainous regions of the western U.S. Enhanced precipitation rates in seeded cloud regions have been measured in the range of hundredths to >1 mm per hour. Although conducted over smaller temporal and spatial scales, research results tend to be consistent with evaluations of randomized experiments and a substantial and growing number of operational programs where 5% to 15% increases in seasonal precipitation have been consistently reported. Similar results have been found in both continental and coastal regions, with the potential for enhanced precipitation in coastal regions appearing to be greater in convective cloud regimes. The consistent range of indicated effects in many regions suggests fairly widespread transferability of the estimated results.

Technological advances have aided winter precipitation augmentation programs. Fast-acting silver iodide ice nuclei, with higher activity at warmer temperatures, have increased the capability to augment precipitation in shallow orographic cloud systems. Numerical modeling has improved the understanding of atmospheric transport processes and allowed simulation of the meteorological and microphysical processes involved in cloud seeding. Improvements in computer and communications systems have resulted in a steady improvement in remotely controlled cloud (ice) nuclei generators (CNG's), which permit improved placement of CNG's in remote mountainous locations.

Wintertime snowfall augmentation programs can use a combination of aircraft and ground-based dispersing systems. Although silver iodide compounds are still the most commonly used glaciogenic (causing the formation of ice) seeding agents, dry ice is used in some warmer (but still supercooled) cloud situations. Liquid propane also shows some promise as a seeding agent when dispensers can be positioned above the freezing level on the upwind slopes of mountains at locations adequately far upwind to allow growth and fallout of precipitation within the intended target areas. Dry ice and liquid propane expand the window of opportunity for seeding over that of silver iodide, since they can produce ice particles at temperatures as warm as −0.5°C. For effective precipitation augmentation, seeding methods and guidelines need to be adapted to regional meteorological and topographical situations.

Although traditional statistical methods continue to be used to evaluate both randomized and non-randomized wintertime precipitation augmentation programs, the results of similar programs are also being pooled objectively in order to obtain more robust estimates of seeding efficacy. Objective evaluations of non-randomized operational programs continue to be a difficult challenge. Some new methods of evaluation using the trace chemical and physical properties of segmented snow profiles show considerable promise as possible means of quantifying precipitation augmentation over basin-sized target areas.

Summer Precipitation Augmentation

The capability to augment summer precipitation from convective clouds has been reasonably well demonstrated. Assessments of some operational and research programs that have seeded selected individual clouds or clusters of clouds with either glaciogenic or hygroscopic nuclei have found that seeded clouds tend to last longer, expand or travel farther to cover larger areas, and are more likely to merge with nearby clouds and produce more precipitation. Both dynamic and microphysical changes appear to be involved.

Results from research programs conducted on summertime cumulus clouds are encouraging but somewhat variable. Part of the resulting uncertainty is due to the variety of climatological and microphysical settings in which experimentation has been conducted. Other important factors include the spatial scale at which the investigations are conducted and the seeding mode. Projects which relied upon introduction of glaciogenic seeding material targeted for specific clouds or portions of clouds that met certain criteria (based essentially upon the stage of development of the clouds) have generally indicated positive seeding effects, ranging between 50% and 100% for individual clouds and on the order of 50% for clusters of convective clouds.

Evaluations of operationally conducted summer precipitation augmentation programs present a difficult problem due to their non-randomized nature and the normally large temporal and spatial variability present in summertime rainfall. Recognizing these evaluation limitations, various methods for the evaluation of such programs have been developed and used, ranging in scale from individual clouds to floating targets of varying sizes to area-wide analyses. The results of many of these evaluations, at the single cloud scale through floating target areas up to 1,700 km² have indicated a positive seeding effect in precipitation. Area-wide effects can be more difficult to discern due to the large temporal and spatial variability in summertime rainfall noted earlier. In some instances, apparent positive effects of seeding have also been noted outside the specific targets. Thus, the apparent effect of seeding is not necessarily confined to the directly treated clouds. The physical mechanisms leading to those effects outside the directly treated clouds are not yet fully understood.

Technological advances have aided summer precipitation augmentation programs. These include fast-acting silver iodide ice nuclei, new hygroscopic seeding formulations, sophisticated radar and satellite data processing and analysis capabilities, airborne cloud physics instrumentation and continued improvements in numerical modeling.

5.4.3 American Meteorological Society

The following paragraphs provide excerpts from the AMS Statement on Planned and Inadvertent Weather Modification (AMS, 1998).

There is growing evidence that glaciogenic seeding (the use of ice-forming materials) can, under certain weather conditions, successfully modify supercooled fog, some orographic stratus clouds, and some convective clouds. Recent research results utilizing both in situ and remote measurements in summer and winter field projects provide dramatic though limited evidence of success in modifying shallow cold orographic clouds and single-cell convective clouds. Field studies are beginning to define the frequencies with which responsive clouds occur within specific meteorological regimes.

Successful treatment of any suitable cloud requires that sufficient quantities of appropriate seeding materials must enter the cloud in a timely, well-targeted fashion. As the need for stringent spatial and temporal targeting has been established, it has become apparent that problems with seeding plume delivery in many early experiments may in part account for the failure of such projects to produce significant results.

Precipitation increase

There is considerable evidence that, under certain conditions, precipitation from supercooled orographic clouds can be increased with existing techniques. Statistical analyses of precipitation records from some long-term projects indicate that seasonal increases on the order of 10% have been realized. The cause-and-effect relationships have not been fully documented; however, the potential for increases of this magnitude is supported by field measurements and numerical model simulations. Both show that supercooled liquid water exists in amounts sufficient to produce the observed precipitation increases and could be tapped if proper seeding technologies were applied. The processes culminating in increased precipitation have recently been directly observed during seeding experiments conducted over limited spatial and temporal domains. While such observations further support statistical analyses, they have to date been of limited scope, and thus the economic impact of the increases cannot be assessed.

Recent experiments continue to suggest that precipitation from single-cell and multicell convective clouds may be increased, decreased, and/or redistributed. The response variability is not fully understood, but appears to be linked to variations in targeting, cloud selection criteria, and assessment methods.

Heavy glaciogenic seeding of some warm-based convective clouds (bases at +10°C or warmer) can stimulate updrafts through added latent heat release (a dynamic effect) and consequently increase precipitation. However, convincing evidence that such seeding can increase rainfall over economically significant areas is not yet available.

Seeding to enhance coalescence or affect other warm rain processes within clouds having summit temperatures warmer than about 0°C has produced statistically acceptable evidence of accelerated precipitation formation within clouds, but evidence of rainfall change at the ground has not been obtained.

Although some present precipitation augmentation efforts are reportedly successful, more consistent results would probably be obtained if some basic improvements in seeding methodology were made. Transport of seeding materials continues to be uncertain, both spatially and temporally. Improved delivery techniques and better understanding of the subsequent transport and dispersion of the seeding materials are needed. Current research using gaseous tracers such as sulfur hexafluoride is addressing these problems.

There are indications that precipitation changes, either increases or decreases, can also occur at some distance beyond intended target areas. Improved quantification of these extended (extra-area) effects is needed to satisfy public concerns and assess hydrologic impacts.

Precipitation augmentation projects are unlikely to achieve higher scientific credibility until more complete understanding of the physical processes responsible for any modification effect is established and linked by direct observation to the specific methodology employed. Continued research emphasizing in situ measurements, atmospheric tracers, a variety of remote sensing techniques, and multidimensional numerical cloud models that employ sophisticated microphysics offer improved prospects that this can be accomplished.

5.4.4 World Meteorological Organization

The following are excerpts from the WMO Statement on Weather Modification (WMO, 1992).

Orographic Clouds

In our present state of knowledge, it is considered that the glaciogenic seeding of clouds or cloud systems either formed, or stimulated in development, by air flowing over mountains offers the best prospects for increasing precipitation in an economically viable manner. These types of clouds attract great interest in modifying them because of their potential in terms of water management, i.e., the possibility of storing water in reservoirs or in the snowpack of higher elevation. Numerous research and operational projects conducted since the beginning of weather modification as a science provide the evidence. Statistical analyses suggest seasonal increases (usually over the winter/spring period) on the order of 10% to 15% in certain project areas.

Physical studies using the new technology highlighted above give convincing evidence of the production of an effective seeding agent, the tracing of the agent to supercooled liquid water portions of the cloud, the initiation and development of ice crystals to precipitation size particles, and the fallout of additional precipitation on the mountain slopes in favorable situations over limited areas. Numerical simulations of the process corroborate the physical studies.

This does not imply that the problem of precipitation enhancement in such situations is solved. Much work remains to be done in pursuit of the goals of strengthening the results and producing incontrovertible statistical and physical evidence that the increases occurred over a wide area, over a prolonged period of time, and with minimum, or positive, extra area effects. Existing methods should be improved in the identification of seeding opportunities and the times and situations in which it is not advisable to seed, thus optimizing the technique and quantifying the results.

Also, it should be recognized that the successful conduct of an experiment or operation is a difficult task that requires competent scientists and operational personnel. It is difficult and expensive to safely fly aircraft in supercooled regions of clouds. Such flying requires experienced crews and aircraft with deicing equipment and sufficient power to carry the heavy ice loads that are sometimes acquired. It is also difficult to target the seeding agent from ground generators or from broad-scale seeding by aircraft upwind of an orographic cloud system.

There is limited physical evidence that deliberate heavy seeding of clouds in certain mountainous situations can result in the diversion of snowfall (up to 50 km). However, seeding trials of this type have not been subjected to statistical or numerical modeling evaluation.

Stratiform Clouds

The seeding of cold stratiform clouds began the modern era of weather modification. Deep stratiform cloud systems (but still with cloud tops warmer than −20°C) associated with cyclones and fronts produce significant amounts of precipitation. A number of field experiments and numerical simulations have shown the presence of supercooled water in some regions of these clouds, and there is accumulating evidence that increased precipitation can be obtained by glaciogenic seeding of such volumes. Shallow stratiform clouds can be made to precipitate, often resulting in clearing skies in the region of seeding. One project using these techniques attempts to allow more sunshine to a city, thus reducing the energy requirements of the metropolitan area. The general applicability of these results—when, where, and how extensive could the seeding be in various regions of the world—has not been determined. A worldwide cloud climatology would be useful for this task as well as others listed in this report.

Cumuliform Clouds

In many regions of the world, cumuliform clouds are the main precipitation producers. Cumulus (from small fair weather cumulus to giant thunderstorms) are characterized by vertical velocities often greater than 1.0 m/s and, consequently, contain high condensation rates. They can contain the largest condensed water contents of all cloud types and can yield the highest precipitation rates. Their strong vertical currents can suspend particles for a long enough time for them to grow to large sizes (hail, large raindrops).

For these reasons, cumulus clouds appear to be candidates for modification according to both the static and the dynamic seeding hypotheses. Field experiments with in-cloud microphysical measurements during experimental seeding trials in several regions have shown that isolated cold cumulus clouds which do not produce rain naturally can be stimulated to produce rain by ice-phase cloud seeding. However, the rainfall amounts from these isolated clouds are very small. Reports of limited success have been obtained from attempts to prove that statistically significant rainfall amounts can be produced on a seasonal basis from these cumuli and larger systems. Attempts to significantly enhance rainfall from cumuliform clouds have concentrated their efforts on systems which produce rainfall naturally.

A long-standing program to augment rainfall from wintertime cumulus in the eastern Mediterranean is one of the most widely accepted examples of precipitation enhancement (13%–15% increases) associated with a seeding experiment. Research and operations continue, with recent results indicating the presence of dust affecting the results in one region in a detrimental fashion. Randomized experiments in seeding of warm-based cumulus congestus associated with raining thunderstorms have demonstrated the possibility of enhancing rainfall from such clouds by intensive seeding. Extending this result to increasing the rain over an area met with difficulties. Other randomized experiments have reported enhancement of rainfall from warm-based multicell thunderstorms; those results are still unclear and under international review. New randomized experiments in rain enhancement are being prepared in several areas.

Enhancement of Rain from Warm Clouds

In most countries, the source of water is precipitation, and in tropical regions that precipitation is generally in the form of convective showers, from clouds with tops often not exceeding the height of the freezing level of the so-called warm clouds. In these clouds, the physical processes involved in the initiation and development of rain are condensation, collision-coalescence, and breakup.

Depending on the environment in which these clouds are formed and developed, mainly the type of cloud condensation nuclei (CCN) distribution made available to the system, the growth of large drops can be sufficiently delayed in such a way that the cloud may dissipate before drops grow to precipitation sizes.

The possibility of affecting the condensation/collision-coalescence/breakup growth processes by seeding the cloud with either a hygroscopic material (e.g., artificial CCN) or with small water drops, therefore tapping the potential precipitation efficiency of the cloud system, has led to the hypothesis of rain enhancement from warm clouds.

Most of the warm rain processes have been simulated both in laboratory as well as in modeling work. Although favorable from the theoretical point of view, the experiments for rain enhancement from warm clouds conducted up to the present time, do not have the necessary physical observations for clear-cut evaluation and possible technology transfer.

5.5 CONCLUSIONS

There are a variety of cloud seeding modes and types of instrumentation available for application in precipitation augmentation seeding projects. The selection of a particular cloud seeding mode or instrument to be used will vary depending on the unique characteristics of a given project area and project goals. Factors affecting these selections include time of operations (winter, summer), target area size, topography, accessibility, funding available, and other project-specific aspects. It is highly desirable to consider these decisions prior to the initiation of an operational cloud seeding project. A project design performed for a particular project should specify cloud seeding modes and instrumentation to be used as well as provide additional information necessary for the conduct of the project.

Specification of a cloud seeding mode includes considerations of the type of cloud seeding agent to be used as well as the type of dispensing technique to be used to disperse the cloud seeding agent. A variety of seeding agents have received varying degrees of attention since 1946. Among the most commonly used agents in operational projects are silver iodide and dry ice. Other agents are available, including liquid propane, but several others suffer from one or more of the following limitations: (1) lower seeding effectiveness than silver iodide or dry ice; (2) some attendant environmental concerns that might be associated with the release of lead iodide; or (3) the lack of demonstrated seeding capability in the atmosphere instead of that indicated solely through laboratory trials (Griffith et al., 1995).

Dispensing systems can be categorized as either for ground-based or aerial use. Ground-based systems are most useful in wintertime projects in mountainous areas; they have limited utility in summertime projects. Aerial dispensing systems are ideally suited to summertime cumulus seeding either at cloud base, in cloud, or at cloud top. Both silver iodide and dry ice can be dispensed aerially; silver iodide and liquid propane can be dispensed from the ground, but dry ice cannot. There are both advantages and disadvantages associated with silver iodide and dry ice as well as ground and aerial dispensing systems. Decisions regarding which to use depend significantly on the requirements and design of a specific project.

Results from research programs conducted in South Africa, Mexico, and Thailand on individual clouds have generated renewed interest in hygroscopic seeding to affect the collision-coalescence processes in clouds or portions of clouds that are greater than 0° C. The viability of this technique to generate increases in precipitation over a fixed target area for a summer season has yet to be established.

Instrumentation for precipitation enhancement projects can serve dual functions: (1) real-time project monitoring; and (2) post-project assessments. Instrumentation, both existing and serving other functions, or installations directly related to the project, provides needed input to real-time decisions, such as the forecasting of probable seeding opportunities, the determination of seedable situations, conduct of seeding operations, and the exercise of project suspension criteria. Instrumentation measurements can also serve in a post-project assessment of the probable effects of cloud seeding based upon critical parameters, such as precipitation or streamflow.

5.6 REFERENCES

Most of the following references were used in Griffith et al. 1995. However, some recent material was needed to update this revision of Manual #81.

American Meteorology Society (1998). "Policy statement: Planned and inadvertent weather modification." *BAMS*, 79, 2771–2772.

American Society of Civil Engineers (2003). Policy Statement #275, Atmospheric Water Management, ASCE, Reston, VA.

American Society of Civil Engineers (2004). *Standard Practice for the Design & Operation of Precipitation Enhancement Projects*, ASCE/EWRI Standard 42-04, ASCE, Reston, VA, 59 p.

Andreae, M.O., Rosenfeld, D., Artaxo, P., Costa, A.A., Frank, G.P., Longo, K.M., and Silva-Dias, M.A.F. (2004). "Smoking rain clouds over the Amazon." *Science*, **303**, 1337–1342.

Appleman, H. (1958). "An Investigation into the Formation of Hail." *Nubila*, 1, 28–37.

Baer, V.E. (1991). "The Transition from the Present Radar Dissemination System to the NEXRAD Information Dissemination System." *Bull. Amer. Meteor. Soc.*, **72**, AMS, Boston, MA, 29–33.

Benner, R.L., and Lamb B. (1985). "A Fast Response Continuous Analyzer for Halogenated Atmospheric Tracers." *J. Atmospheric Oceanic Technology*, **2**, AMS, Boston, MA, 582–589.

Ben-Zvi, A. (1988). "Enhancement of runoff from a small watershed by cloud seeding." *J. Hydrology*, **101**, 291–303.

Ben-Zvi, A. (1989). "Projection onto ground rainfall distributions of effects of seeding convective cloud cells." *J. Weather Modification*, **21**, 62–64.

Ben-Zvi, A., and Fanar A. (1996). "Effects of cloud seeding on rainfall intensities in Israel." *Israel J. Earth Sciences*, **45**, 39–54.

Ben-Zvi, A., and Langerman M. (1989). "Estimation of flow enhancement in the Kinneret watershed due to cloud seeding." *Stochastic Hydrology & Hydraulics*, **3**, 203–216.

Ben-Zvi, A., and Langerman M. (1995). "Assessment of streamflow augmentation in case of weakly dependent data." *J. Irrigation Drainage Engineering*, **121**, ASCE, Reston, VA, 104–108.

Boe, B.A., Stith, J.L., Smith, P.L., Hirsch, J.H., Helsdon, Jr., J.H., Detwiler, A.G., Orville, H.D., Martner, B.E., Reinking, R.F., Meitin, R.J., and Brown, R.A. (1992).

"The North Dakota Thunderstorm Project: A cooperative study of High Plains thunderstorms." *Bull. Amer. Meteor. Soc.*, **73,** 145–160.

Brown, K. J., Thompson, J.R., St. Amand, P., and Elliott, Jr., S.D. (1974). "The Seeding of Convective Bands." *Fourth Conference on Weather Modification*, AMS, Boston, MA, 7–12.

Bruintjes, R.T., Breed, D.W., Foote, G.B., Dixon, M.J., Brown, B.G., Salazar, V., and Rodriguez, H.R. (1999). "Program for the augmentation of rainfall in Coahilla (PARC): Overview and design." *Preprints, Seventh WMO Scientific Conf. Wea. Modif.*, Technical Document No. 936, World Meteorology Organization, Geneva, 53–56.

Chai, S.K, Finnegan, W.G., and Pitter, R.L. (1993). "An interpretation of the mechanisms of ice-crystal formation operative in the Lake Almanor Cloud Seeding Program." *J. Appl. Meteorology*, **32**, AMS, Boston, MA, 1726–1732.

Changnon, S.A., Huff, F.A., and Semonin, R.G. (1971). "Metromex: An Investigation of Inadvertent Weather Modification." *Bull. Amer. Meteor. Soc.*, **52,** MAS, Boston, MA, 958–967.

Church, J.E. (1918). *Snow Surveying*. Western Engineering.

Cooper, C.F., and Jolly W.C. (1969). Ecological Effects of Weather Modification. University of Michigan, School of Natural Resources, Report to the Department of the Interior.

Cooper, W.A., Bruintjes, R.T., and Mather, G.K. (1997). "Some calculations pertaining to hygroscopic seeding with flares." *J. Appl. Meteorology*, **36**, AMS, Boston, MA, 1449–1469.

DeFelice, T.P. (1998). *An Introduction to Meteorological Instrumentation and Measurement.* Prentice Hall, NJ, 229 pp.

DeMott, P.J. (1988). "Comparisons of the Behavior of AgI-Type Ice Nucleating Aerosols in Laboratory Simulated Clouds." *J. Weather Modification*, **20**, WMA, Fresno, CA, 44–50.

DeMott, P.J., Finnegan, W.G., and Grant, L.O. (1983). Ecological Effects of Weather Modification. University of Michigan, School of Natural Resources, Report to the Department of the Interior.

Dennis, A.S. (1980). "Weather Modification by Cloud Seeding." International Geophysics Series, **24**, Academic Press, New York, NY, 24.

Deshler, T. (1988). "Corrections of Surface Particle Probe Measurements for the Effects of Aspiration." *J. Atmospheric Oceanic Technology*, **5**, AMS, Boston, MA, 547–560.

Deshler, T., Reynolds, D.W., and Huggins A.W. (1990). "Physical Response of Winter Orographic Clouds over the Sierra Nevada to Airborne Seeding Using Dry Ice or Silver Iodide." *J. Appl. Meteorology*, **29**, AMS, Boston, MA, 288–330.

Dixon, M., and Wiener, G. (1993). "TITAN: Thunderstorm Identification, Tracking, Analysis and Nowcasting—A radar-based methodology." *J. Atmospheric Oceanic Technology*, **10**, 785–797.

Donnan, J.A., Blair, D.N., Finnegan, W.G., and St. Amand, P. (1970). "Nucleation Effectiveness of AgI-NH_4I and AgI-NaI Acetone Solutions and Pyrotechnic Generators as a Function of LWC and Generator Flame Temperature." *J. Weather Modification*, **2**, WMA, Fresno, CA, 155–164.

Douglas, W. J. (1968). Toxic Properties of Material Used in Weather Modification. *Proceedings of the First National Conference on Weather Modification*, State University of New York, Albany, NY.

Finnegan, W.G. (1999). "Generation of Ice Nucleus Aerosols by Solution and Pyrotechnic Combustion." *J. Weather Modification*, **31**, AMS, Fresno, CA, 102–108.

Finnegan, W.G., and Pitter, R.L. (1987). "Rapid Ice Nucleation by Acetone-Silver Iodide Generator Aerosols." *J. Weather Modification*, **20**, WMA, Fresno, CA, 51–53.

Fukuta, N. (1963). "Ice Nucleation by Metaldehyde." *Nature*, **199**, London, England, 475–476.

Fukuta, N. (1966). "Experimental Studies of Organic Ice Nuclei." *J. Atmospheric Science*, **23**, AMS, Boston, MA, 191–196.

Fukuta, N., Plooster, M.N., Armstrong, J.A., Gorore, A., Butz, J.R., Schaller, R.C., and Glen, L. (1977). The Development of Organic Ice Nuclei Generators for Weather Modification. Final Report, Denver Research Institute, Report to the National Science Foundation, 2 Volumes.

Gagin, A., and Gabriel, K.R. (1987). "Analysis of recording rain gauge data for the Israeli II experiment; Part 1: Effect of cloud seeding on the components of daily rainfall." *J. Climatology Appl. Meteorology*, **20**, AMS, Boston, MA, 1301–1321.

Garvey, D.M. (1975). "Testing of Cloud Seeding Materials at the Cloud Simulation and Aerosol Laboratory." *J. Appl. Meteorology*, **14**, AMS, Boston, MA, 883–890.

Givati, A., and Rosenfeld, D. (2004). "Quantifying precipitation suppression due to air pollution." *J. Appl. Meteorology*, **43**, AMS, Boston, MA, 1038–1056.

Givati, A., and Rosenfeld, D. (2005). "Separation between Cloud Seeding and Air Pollution Effects." (In press.) *J. Appl. Meteorology*, **44.**

Griffith, D.A., Wilkerson, G.W., and Risch, D.A. (1990). "Airborne Observations of a Summertime, Ground-Based Tracer Gas Release." *J. Weather Modification*, **22**, WMA, Fresno, CA, 43–48.

Griffith, D.A., Wilkerson, G.W., Hauze, W.J., and Risch, D.A. (1992). "Observations of Ground Released Sulfur Hexafluoride Tracer Gas Plumes in Two Utah Winter Storms." *J. Weather Modification*, **24**, WMA, Fresno, CA, 49–65.

Griffith, D. A., Politovich, M., Renick, J.H., Reynolds, D.W., and Rogers, D. (1995). "Cloud Seeding Modes and Instrumentation." ASCE Manual of Professional Practice No. 81, Section 5, *Guidelines for Cloud Seeding to Augment Precipitation*, ASCE, Reston, VA, pp. 87–123.

Guillory, A.R., Jedlovec, G.J., and Fuelberg, H.E. (1993). "A Technique for Deriving Column-Integrated Water Content Using VAS Split-Window Data." *J. Appl. Meteorology*, **32**, AMS, Boston, MA, 1226–1241.

Hastay, M., and Gladwell, J.S. (1969). "Statistical Evaluations of a Cloud Seeding Project at the Streamflow Control Level." *J. Hydrology*, North Holland Publishing Company, Amsterdam, the Netherlands, 117–135.

Henderson, T.J. (1981). "Weather Resources Management on the Kings River Watershed." *J. Weather Modification*, **13**, (1), WMA, Fresno, CA, 121–131.

Hindman, E.E. (1986). "Characteristics of Supercooled Liquid Water Clouds at Mountaintop Sites in the Colorado Rockies." *J. Climate Appl. Meteorology*, **25**, AMS, Boston, MA, 1271–1279.

Hogg, D.C., Guirard, F.O., Snider, J.B., Decker, M.T., and Westwater, E.R. (1983a). "A Steerable Dual-Channel Microwave Radiometer for Measurement of Water Vapor and Liquid in the Troposphere." *J. Appl. Meteorology*, **22**, AMS, Boston, MA, 789–806.

Hogg, D.C., Decker, M.T., Guirard, F.O., Earnshaw, K.B., Merritt, D.A., Moran, K.P., Sweezy, W.B., Strauch, R.G., and Westwater, E.R. (1983b). "An Automatic Profiler of the Temperature, Wind and Humidity in the Atmosphere." *J. Climate Appl. Meteorology*, **22**, AMS, Boston, MA, 807–831.

Holroyd, E.W., Super, A.B., and Silvermann, B.A. (1978). "The Practicability of Dry Ice for on-Top Seeding of Convective Clouds." *J. Appl. Meteorology*, **17**, AMS, Boston, MA, 49–63.

Horn, R.D., Finnegan, W.G., and DeMott, P.J. (1983). "Experimental Studies of Nucleation of Dry Ice." *J. Appl. Meteorology*, **22**, AMS, Boston, MA, 1567–1570.

Klazura, G.E. (1993). "A Description of the Initial Set of Analysis Products Available from the NEXRAD WSR-88D System." *Bull. Amer. Meteor. Soc.*, **74**, AMS, Boston, MA, 1293–1311.

Klazura, G.E., Thomale, J.M., Kelly, D.S., and Jendrowski, P. (1999). "A comparison of NEXRAD WSR-88D radar estimates of rain accumulation with gauge measurements for high- and low-reflectivity horizontal gradient precipitation events." *J. Atmos. Oceanic Tech.*, **16,** 1842–1850.

Klein, D.A. (1978). Management of Nucleating Agents Used in Weather Modification. Final Report Submitted to National Science Foundation, Project ENV73–07821A05.

Kumai, M. (1982). "Formation of Ice Crystals and Dissipation of Supercooled Fog by Artificial Nucleation, and Variations of Crystal Habit of Early Growth Stages." *J. Appl. Meteorology*, **21**, AMS, Boston, MA, 579–587.

Langer, G. (1973). "Evaluation of NCAR Ice Nucleus Counter. Part 1: Basic Operation." *J. Appl. Meteorology*, **12**, AMS, Boston, MA, 1000–1011.

Langer, G., Rosinski, J., and Bernsen, S. (1963). "Organic Crystals as Icing Nuclei." *J. Atmospheric Science*, **20**, AMS, Boston, MA, 557–562.

Ludlam, F.H. (1958). "The Hail Problem." *Nubila*, 1, 12–95.

Martner, B.E., and Kropfli, R.A. (1989). "Tracir: a Radar Technique for Observing the Exchange of Air Between Clouds and Their Environment." *Atmospheric Environment*, **23**, AMS, Boston, MA, 2715–2721.

Mather, G.K., and Terblanche, D. (1994). Initial results from cloud seeding experiments using hygroscopic flares. *Preprints, Sixth WMO Scientific Conference on Weather Modification*, WMP Report No. 22, World Meteorology Organization, Geneva, 687–690.

Mather, G.K., Dixon, M.J., and deJager, J.M. (1996). "Assessing the potential for rain augmentation—the Nelspruit randomized convective cloud seeding experiment." *J. Appl. Meteorology*, **35**, AMS, Boston, MA, 1465–1482.

Mather, G.K., Terblanche, D.E., Steffens, F.E., and Fletcher, L. (1997). "Results of the South African cloud-seeding experiments using hygroscopic flares." *J. Appl. Meteorology*, **36**, 1433–1447.

Mossop, S.C. (1955). The Freezing of Supercooled Water. *Proceedings of the Physics Society*, **68**, 165–174.

Pasqualucci, F., Bartram, B.W., Kropfli, R.A., and Moninger, W.R. (1983). "A Millimeter Wavelength Dual Polarization Doppler Radar for Cloud and Precipitation Studies." *J. Appl. Meteorology*, **22**, AMS, Boston, MA, 758–765.

Power, B.A., and Power, R.F. (1962). "Some Amino Acids as Ice Nucleators." *Nature*, **194**, London, England, 1170–1171.

Price, G.A., and Rilling, R.A. (1987). A High Resolution Infinite Capacity Weighing Snow Gauge for Use in Remote Areas. *Preprints, Sixth Symposium on Meteorological Observations and Instrumentation*, New Orleans, AMS, Boston, MA.

Ramanathan, V., Crutzen, P.J., Kiehl, J.T., and Rosenfeld, D. (2001). "Aerosols, Climate and the Hydrological Cycle." *Science*, 294, 2119–2124.

Reynolds, D.W. (1988). "A Report on Winter Snowpack Augmentation." *Bull. Amer. Meteor. Soc.*, AMS, Boston, MA, 1290–1300.

Reynolds, D.W. (1991). "Design and Field Testing of a Remote Ground-Based Liquid Propane Dispenser." *J. Weather Modification*, **22,** 49–53.

Reynolds, D.W. (1992). "A Snowpack Augmentation Program using Liquid Propane." *Preprints, Symposium on Planned and Inadvertent Weather Modification*, AMS, Boston, MA, 88–95.

Reynolds, D.W., and Dennis, A.S. (1986). "A Review of the Sierra Cooperative Pilot Project." *BAMS*, 67, 513–523.

Rosenfeld, D. (1999). "TRMM Observed First Direct Evidence of Smoke from Forest Fires Inhibiting Rainfall." *Geophysical Research Letters*. **26,** (20), 3105–3108.

Rosenfeld, D. (2000). "Suppression of Rain and Snow by Urban and Industrial Air Pollution." *Science*, **287** (5459), 1793–1796.

Rosenfeld, D. and Lensky, I.M., (1998). "Satellite-based insights into precipitation formation processes in continental and maritime convective clouds." *Bull. Amer. Meteor. Soc.*, **79**, 2457–2476.

Rosenfeld, D., Lahav, R., Khain, A. P., Pinsky, M. (2002). The role of sea-spray in cleansing air pollution over ocean via cloud processes. "Research Article" in *Science*, **297**, 1667–1670. Highlighted and published online in "*Science Express*" on Aug. 15, 2002.

Rudich, Y., Rosenfeld, D., Khersonsky, O. (2002). "Treating clouds with a grain of salt." *Geophysical Research Letters*. **29** (22), doi:10.1029/2002GL016055, 2002.

Sassen, K. (1992). "The Polarization Lidar Technique for Cloud Research: A Review and Current Assessment." *Bull. Amer. Meteor. Soc.*, **12**, AMS, Boston, MA, 1848–1866.

Schaefer, V.H. (1946). "The Production of Ice Crystals in a Cloud of Supercooled Water Droplets." *Science*, **104**, (2707), 459.

Sharon, D. (1978). "Rainfall fields in Israel and Jordan and the effect of cloud seeding on them." *J. Appl. Meteorology*, **17**, 40–48.

Stith, J.L., Griffith, D.A., Rose, R.L., Flueck, J.A., Miller, Jr., J.R., and Smith, P.L. (1986). "Aircraft Observations of Transport And Diffusion in Cumulus Clouds." *J. Climate Appl. Meteorology*, 25, AMS, Boston, MA, 1959–1970.

Stone, R.H., and Marler, B.L. (1993). Phase III—The Use of Trace Chemistry for the Assessment of Cloud Seeding Effect in the Lake Almanor Watershed. Pacific Gas and Electric Company, San Francisco, CA.

Super, A.B. (1991). "Winter Orographic Cloud Seeding Status in the Intermountain West." *J. Weather Modification*, 22, WMA, Fresno, CA, 106–116.

Super, A.B., and Huggins, A.W. (1992). "Investigations of the targeting of ground-released silver iodide in Utah—Part I: Ground observations of silver-in-snow and ice nuclei." *J. Weather Modification*, **24,** 19–34.

Super, A.B., and Holroyd, III, E.W. (1998). Snow accumulation algorithm for the WSR-88D radar: Final report. Report No. 98-05, Bur. Rec., Technical Services Center, Denver, CO, 75 pp.

Super, A.B. (1999). "Summary of the NOAA/Utah Atmospheric Modification Program: 1990-1998." *J. Weather Modification,* 31, WMA, Fresno, CA, 51–75.

Vickers, W.W., and Church, J.F. (1966). "Investigation of Optimal Design for Supercooled Cloud Dispersal Equipment and Techniques." *J. Appl. Meteorology,* **5**, AMS, Boston, MA, 105–118.

Vonnegut, B. (1947). "The Nucleation of Ice Formation by Silver Iodide." *J. Applied Physics,* 18, AMS, Boston, MA, 593–595.

Warburton, J.A. (1974). Physical Evidence of Transport of Cloud Seeding Materials into Areas Outside Primary Targets. Reprints of International Conference on Weather Modification, Canberra, Australia, 185–190.

Warburton, J.A., and DeFelice, T.P. (1986). "Oxygen isotopic composition of central Sierra Nevada precipitation. I.: Identification of ice phase water capture regions in winter storms." *Atmospheric Research,* **20**, 11–22.

Warburton, J.A., Chai, S.K., Stone, R.H., and Young, L.G. (1996). "The assessment of snowpack enhancement by silver iodide cloud-seeding using the physics and chemistry of the snowfall." *J. Weather Modification,* **28**, WMA , Fresno, CA, 19–28.

Ward, P.J., and Demott, P.J. (1989). "Preliminary Experimental Evaluation of Snowmax Snow Inducer, *Pseudomonas syringe*, as An Artificial Ice Nucleus for Weather Modification." *J. Weather Modification,* **21**, WMA, Fresno, CA, 9–13.

Weber, B.L., and Wurtz, D.B. (1990). "Comparison of Rawinsonde and Wind Profiler Radar Measurements." *J. Atmos. Oceanic Tech.,* **7**, AMS, Boston, MA, 157–174.

Woodley, W.L, Rosenfeld, D., and Strautins, A. (2000). "Identification of a seeding signature in Texas using multi-spectral satellite imagery." *J. Weather Modification,* **32**, 37–52.

Weather Modification Association (2005). "Weather Modification Association Capability Statement." http://www.weathermodification.org.

World Meteorological Organization (1992). "Statement on the status of weather modification." *J. Weather Modification,* **25**, WMA, Fresno, CA, 1–6.

SECTION 6

HOW TO IMPLEMENT A CLOUD SEEDING PROGRAM

Bruce A. Boe[13], *A.M. ASCE, WMA CM/CO and Conrad G. Keyes Jr.*[14], *P.E., P.S., ScD, Hon. M. ASCE (Life), F. NSPE (Life), D.WRE, WMA CM/CO*

6.1 INTRODUCTION

The common program design elements underlying the different programmatic aspects of a cloud seeding program necessitate that water resource managers collaborate with public officials, meteorologists, environmental scientists, and the local populace when designing operational cloud seeding programs. Local climatology, hydrology, water storage facilities, and environmental concerns must be viewed as a single interconnected whole. The hydrologist quantifying streamflows, the environmental scientist assessing the impacts of precipitation changes and seeding agents, the meteorologist studying precipitation patterns, and the biologist monitoring wildlife and vegetation within and beyond the program area should be consulted when creating the cloud seeding program plan (Section 6.3.2), to ensure that the evaluation portion of this plan (Section 6.4.2) is comprehensive enough to adequately determine the success of the program and to help future researchers evaluate changes potentially associated with the cloud seeding program. Such a team may provide the best basis for determining the long-term future of each program (See Section 2.2). Given a well-constructed plan, the next step is to secure the resources it stipulates, and get them ready for the implementation phase. The implementation of a cloud seeding program, regardless of size, is not as easy as it may sound, and should begin with an all-hands

---

[13] Director of Meteorology, Weather Modification, Inc., 3802 20th Street North, Fargo, ND 58102.

[14] Environmental & Water Resources Consultant and Chair, EWRI Standards Development Council, P.O. Box 1499, Mesilla Park, NM 88047-1499.

meeting. This meeting lays out the role and responsibilities of each resource, project team members, as well as program objectives, initial program assessment or program definition, factors governing implementation, program control, and program management. The sections below provide more detail on how to implement a cloud seeding program. Armed with this information, a prospective cloud seeding sponsor can decide what elements of the process to assume individually or institutionally and whether/when to enlist the services of an expert weather modification consultant.

6.1.1 Initial Program Assessment (Feasibility Study)

Before a cloud seeding program is implemented, a feasibility study (or initial program assessment), as described in Section 6.3, should be conducted to assess the probability of the program becoming successful. Should the feasibility study indicate that the program (e.g., cold cloud precipitation augmentation, warm cloud precipitation augmentation) would likely be unsuccessful, there should be no implementation phase for that type of program.

6.1.2 The Factors Governing Implementation

Cloud seeding programs to increase precipitation are implemented primarily because there is a need for additional water. The need and initial program assessment (or feasibility study) help define how much additional water is desired, and how likely the existing technology and its science will ease the water shortfall, without creating problems from excess water. Thus, a goal is defined and the steps of the planning and implementation should follow as provided in Figure 6-1.

The feasibility study (Section 6.3) determines the amount of additional water that can be expected from a program. It considers the local climatology, the seeding technology(ies) to be applied, and any known operational constraints that may exist.

If the findings from the feasibility study reveal no difficulties, the program can then be clearly defined. Included with the program design should be a plan for program evaluation. Once the program is defined, the needs and goals could be refined. The program design, in addition to maximization of economic benefits, must protect the public interest and safety and avoid (or minimize) adverse impacts. Recreation may be improved through increased water in lakes, reservoirs, and/or more snowpack. There should also be an evaluation plan (assessment) to quantify the results of the program in terms appropriate to the program sponsor's needs. A program evaluation (assessment) is very important to the longevity of any program.

Program controls are then set in place defining the program infrastructure. Program controls, as shown in Figure 6-1, imply communications (flow of information) and the criteria for seeding and for suspending seeding. Program controls are both scientific and political, the latter

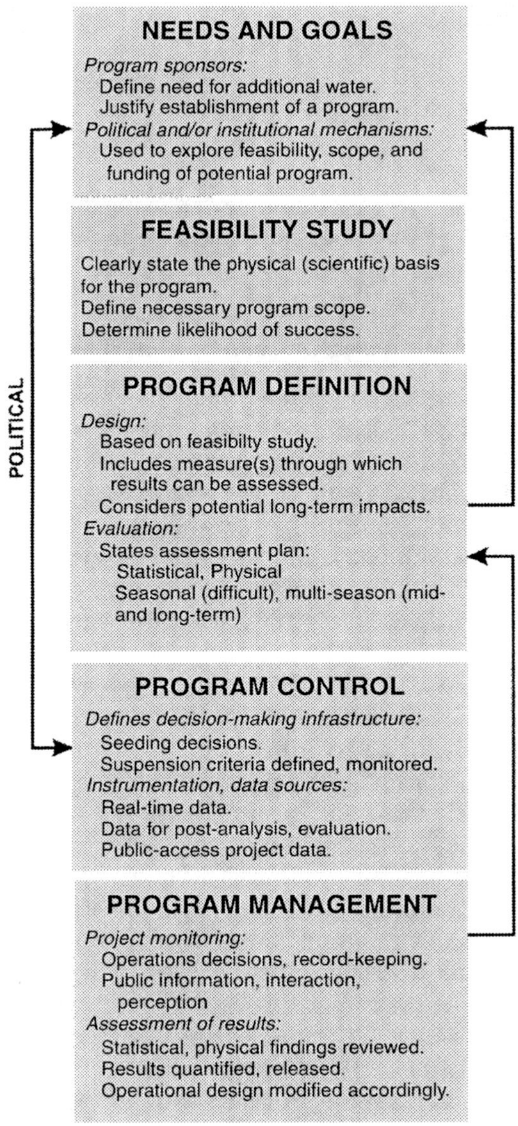

FIGURE 6-1. The Steps Involved in the Planning and Implementation of a Cloud Seeding Program.

because it is usually the will of the people (politics) that leads to the creation of programs. Program controls affect program goals. For example, suspension criteria may necessarily be conservative, which is then reflected in the stated program goals.

Program management has direct input in the program design and evaluations, but is also responsible for other essential functions, including public relations, assessment, record keeping, and day-to-day operations (Section 6.6).

6.2 NEEDS AND GOALS

The primary objective of cloud seeding programs designed to increase precipitation is to help meet the water resource needs of society. This section describes how program sponsors might approach the assessment of need and the formulation of goals.

6.2.1 Origin of Need and Program Justification (Program Sponsors)

Humanity's existing water development technologies have resulted directly from need, and humankind's ingenuity to anticipate and act to meet those needs. Cloud seeding (weather modification) is but one case in point. Society must locally assess all the alternatives for development of water in their area, and determine which will be implemented. Common water resource development tools include conservation, reuse, dam, reservoir, and irrigation system construction, inter-basin diversion, and weather modification. These tools may be applied singly, or in almost any combination.

Social needs are first articulated by those in need, and awareness grows. With awareness comes dialogue and discussion, as may be heard when people speak publicly, or as may be found in the printed media. Such needs are sometimes formally addressed through legislation. Most governments are responsive to their constituencies, and will develop weather modification programs to address expressed water development needs, if the government leaders are convinced of the feasibility of the development of a cloud seeding project. In many (if not most) instances, the initial impetus for a prospective program comes directly from the water resource community (e.g., water districts, hydroelectric utilities, etc.).

The identified needs must then be translated into achievable engineering and scientific objectives. In other words, program viability must be assessed, and the costs of such programs estimated. If the potential sponsors can afford to conduct the program and the program benefits are projected to significantly exceed the costs, then a desire for program implementation usually follows. The professionals collect and analyze the requisite weather, climate, operational aspects, and funding information.

In so doing, the literature searches should include engineering, meteorological, social, environmental, and legal publications on weather modification technology. The findings collectively help determine the final decision to proceed with a feasibility, scope, and funding of the potential program.

6.2.2 Political and/or Institutional Mechanisms

Water resource engineers, meteorologists, economists, and their respective organizations must recognize the need for an interdisciplinary approach to decision making, and apply this approach beginning with the earliest stages of program design and development.

In many cases state and federal laws will place restrictions and/or reporting requirements on proposed programs. Public meetings are often required. Governing agencies must be identified, and those offices having jurisdiction in the program area contacted. If laws exist, it is likely that licenses and/or permits may be required. Complete descriptions of the program, methodologies, seeding agents, and safety criteria are usually required as part of the licensing/permitting process. Considerably more information in this regard is provided in Section 3.

In some cases it may be helpful to establish a citizens' advisory committee, composed of representatives of the area's stakeholder groups. Such a committee would normally represent the major economic interests of the region, local governmental officials (people not involved in program regulation, such as city council members), environmental and public interest groups, and perhaps the news media.

6.3 THE FEASIBILITY STUDY

Planning has been defined as the orderly consideration of a program from the original statement of purpose through the evaluation of alternatives, to the final decision on a course of action (Linsley and Franzini, 1964). There is no predefined process that always leads to the "best" decision, because each cloud seeding program is unique in its physical and financial setting. There is no substitute for professional judgment in program planning, design, and management. Each individual step toward a final program design should be supported whenever possible by quality quantitative analyses rather than estimates.

6.3.1 Scientific Basis

The term "feasibility study" refers to the examination of the local climate and cloud characteristics to determine whether or not cloud seeding technology has a reasonable expectation of increasing precipitation. The term "program assessment" refers to the evaluation of the program itself when it is actually conducted. The program assessment may include operational decision-making procedures, forecasting, and, most often, the effects of the program on precipitation.

The feasibility of a program depends largely upon two factors. First is there a scientific basis for the work proposed that could yield the desired additional precipitation? This is discussed in detail in Section 4. Secondly, even if such a basis exists, is the cost of implementing a program based on the known science affordable? The latter depends heavily upon the combination of available financial resources and the expected return in additional water, in other words, the benefit/cost ratio.

When possible, the feasibility study for a program should draw significantly from previous research and well-conducted operational programs that are similar in nature to the proposed program (e.g., similar topography, similar precipitation occurrences, etc.). Percentage increases obtained from such programs can be used in the development of benefit/cost analysis for the proposed program (see Section 2.4).

6.3.2 Feasibility Study Objectives (Program Scope)

The primary purpose of the feasibility study is to answer two questions. First, does it appear that a cloud seeding program could be implemented in the intended target area that would be successful in achieving the stated objectives of the program. Examples include the increase of high elevation snowpack, and the increase of summer rainfall directly on croplands. Secondly, is the proposed program design expected to produce a positive benefit/cost ratio? The answers to these two questions will determine whether the proposed program appears to be technically and economically feasible.

Answering the first question involves assessing whether or not the climate and cloud characteristics of the region in interest will normally produce sufficient numbers of clouds amenable to effective treatment. In "normal" seasons there must be enough suitable clouds to make a program worthwhile. That number depends upon the increase in precipitation likely to be obtained from each event, and on the value of the additional water thus reaching the surface. Background climatological studies of the weather typical of the intended target area(s) can help address these questions. In addition, the clouds must be treatable; that is, there must be a means of consistently treating the cloud volumes with enough seeding agent(s) to achieve the desired effects. Contributing factors include how seeding agents are transported and dispersed by the airflow and/or convection relative to the locations of the seedable clouds. For orographic seeding, the transport and dispersion is primarily studied relative to the terrain. For airborne seeding, the location(s) of aircraft base(s), the aircraft performance, and the locations climatologically favored for the development of suitable clouds are the primary considerations.

The feasibility study should also address other potential concerns, e.g., the environmental effects of seeding agents such as silver iodide (AgI), and the possibility of measurable downwind effects. References to all relevant research should be summarized for the benefit of the potential program sponsors, and the public. Numerous studies have shown repeatedly that adverse effects are unlikely even with long-term programs. However, if financial resources allow, background measurements of the silver concentrations in the water and soil within and adjacent to the proposed program area, made well before any seeding is ever done, will unequivocally establish the natural background silver levels. Doing so eliminates the possibility of people blaming the seeding program for what they wrongly perceive as elevated silver (or other seeding chemical) levels.

Data often useful in background climatological studies, as detailed in Section 5, include the following:

1. *Precipitation data:* Time-resolved precipitation data are extremely helpful. Daily totals are useful; hourly data even better. If one knows what time precipitation fell, one can correctly correlate the event with the weather conditions at the time. Data can be in water equivalents, as from precipitation gauges, or in snow depth, as from snow boards. Snow pillow data are also useful, but there is a time lag between snowfall and response of the sensors in the pillows of which the analyst must be aware.
2. *Temperature:* Temperatures at the surface and aloft are very useful, for supercooled cloud is necessary for glaciogenic seeding to be effective. Again, time resolved data are preferred; however, upper air soundings are normally made just twice daily and then usually only from fixed locations. Greater time resolution of upper air temperatures can be obtained from prognostic numerical models, but one must remember that such estimated temperatures are from atmospheric models, not actual measurements.
3. *Winds:* Speed and direction, both surface and aloft, are very helpful. In programs proposing to treat orographic clouds (clouds produced by the flow of moist air over hills or mountains), the wind direction is usually strongly correlated with storm conditions, most always when the mid-level horizontal winds are close to perpendicular to the axis of the hills or mountain range. In programs planning to seed convective clouds, the wind speed and direction play a large role in cloud motion.
4. *Humidity:* Measurements at the surface and aloft are very useful, for drier air masses will not produce suitable clouds. Once again, time-resolved data are preferred; however, unless data from

ground-based instrumentation sited on mountains in the area of interest are available, humidity data aloft may be limited to extrapolations from the twice-daily soundings, if available.

5. *Satellite and radar data:* In many locations, satellite and/or radar data may be available. Satellite imagery may offer verification of cloud extent and temperature (at cloud top with infrared imagery) during storm periods. Radar data may likewise be helpful, especially during warmer seasons when the precipitation is not primarily snow.
6. *Streamflow data:* If long-term streamflow data are available, it may be possible to establish correlations between streamflow and snowpack or precipitation (Stauffer, 2001). Such correlations not only are useful in derivation of estimates of streamflow increases, but also may be helpful in designing program safety criteria.

One possible source of estimates of potential program impacts on precipitation is previous research and operational programs conducted in similar climatological settings using the same seeding techniques and seeding agents proposed for the new program. Some examples are the Climax I and II research experiments (Mielke et al., 1981) conducted in the Central Colorado Rockies, the Bridger Range Experiment conducted in southwest Montana (Heimbach and Super, 1988; Super and Heimbach, 1983; Super, 1986), and the High Plains and Edwards Aquifer convective cloud seeding programs of West Texas (Woodley and Rosenfeld, 2004).

If climatological studies provide ample evidence for the existence of suitable clouds, the feasibility study must then address the means through which seeding agent is best delivered to the clouds. This task is commonly referred to as "targeting." The importance of proper targeting cannot be overstated.

There are two means to deliver seeding agents to those clouds deemed amenable to treatment. One is with aircraft, the other from ground-based facilities. Both techniques, outlined below and described in more detail in Section 5, can be effective, but both also have limitations.

1. *Airborne seeding:* Seeding can be done from aircraft in several ways. The primary advantage of airborne seeding is the flexibility of the targeting; seeding agent(s) can be released at almost any coordinates, subject to the restrictions imposed by aircraft performance and the underlying terrain. In winter orographic programs, these restrictions can be significant.
2. *Ground-based seeding:* This can be done from fixed sites, or from mobile platforms, the latter's motion usually being restricted to roadways. The primary advantage of ground-based seeding is that releases can be more or less continuous from a fixed location, such

that a "plume" of seeding agent is consistently transported and dispersed by the wind and air motions to locations downwind. To achieve full coverage of an area, multiple sites are usually required.

Air motions at and near the surface are generally complex, especially near and over hills and mountains, or when convection (thermally driven vertical air motion) is present. Therefore, it is not easy to predict where the seeding agent will be transported, especially given the wide range of possible environmental wind and temperature (stability) profiles. This task could be investigated by three-dimensional numerical modeling, although it is important that the output of the model to be used has been subjected to or is subjected to independent verification (e.g., atmospheric tracer studies). Though not inexpensive, both may impart greater confidence in the targeting strategies being used operationally. Such verifications need not continue indefinitely, but rather only long enough to gain the needed evidence. A variety of models are available, from relatively simple airflow models that can be run on a personal computer to more sophisticated models such as MM5 (Mesoscale Meteorology Model Version 5) and the WRF (Weather Research and Forecasting Model) that require greater effort and a workstation or workstation cluster. The real-time utility of the use of such models to assist in operational decision making during the actual conduct of the program may be limited if several hours are required to perform a run on a supercomputer.

Climatological analyses of available rawinsonde data may also be conducted to determine the mean wind, stability, and vertical temperature profiles that accompany the development of seedable clouds. Hourly observations of available surface wind data may be examined to determine when up-barrier wind components are present. Such information can be used to aid the siting of ground generators in winter programs, and to identify the primary moisture advection regimes for convective programs. For example, a climatological analysis for a winter orographic program might indicate that the predominant wind flow during a majority of seedable situations is associated with mid-level winds from the south through southwest. The analysis might also indicate that these conditions are frequently accompanied by low-level inversions at around the 800 hPa level. Such information would suggest that an array of ground generators should be established south through southwest of the proposed target area, at elevations above 800 hPa. In other words, agent dispensed from generators sited using this information would be expected to have up-barrier trajectories over the intended target area since the seeding material would typically be released above existing stable layers in the lower atmosphere in winds that are heading over the target area.

If the above analyses indicate that a cloud seeding program appears to be technically feasible, then the second question can be answered: Would the program be expected to be economically feasible? In order to answer this question, the value of the expected increase in precipitation must be estimated. For example, in the case of a program to enhance hydropower production, the estimated increase in precipitation will need to be converted to an estimate of increases in streamflow. These estimates of increases in streamflow can be converted into an estimated dollar value, usually drawing upon the potential sponsor's knowledge of the impact of the estimated streamflow on power generation, assuming all additional streamflow can be captured in existing water storage facilities (reservoirs). Program sponsors may well have computer models that can be used to estimate these impacts. Griffith and Solak (2002) provided an example of this type of economic analysis.

The cost of the planned program then needs to be estimated. This may begin as a preliminary estimate that is later refined, based upon the outcome of the design of the proposed program (see Section 6.4). A preliminary benefit/cost ratio may then be calculated. Program sponsors may expect a favorable ratio of perhaps 5:1 or even 10:1 in order for the program to be considered economically feasible, but this depends greatly on the value of water where the program is to be conducted. In areas where an acre-foot of water is worth $1,000, the benefit/cost ratio can be much less and the program can still be very cost effective.

If the initial program design is not found to be cost effective, program designers and sponsors may wish to consider revisions to the initial program design. If such is the case, two questions need to be addressed.

First, what components (if any) of the initial design may not be essential to still have a reasonable probability of achieving the intended goals of the program? Should the initial estimated benefit/cost ratio not be favorable, program designers might ask, "Are there components of the design that are not absolutely essential to achieving the goal(s) of the program?" In other words are there elements that might be "nice to have" but not essential? One way to address this question is to start with a basic core program design, and then rank any potential additions to the program in terms of their perceived ability to "deliver additional precipitation on the ground." A point of diminishing returns may be discovered when this type of analysis is applied, and the program then redesigned accordingly (see Section 2.2.6 and Table 2-1 and 2-2). Once the appropriate adjustments to estimated program costs have been made, the benefit/cost ratio can be recalculated. If a favorable ratio is then obtained, the steps outlined in the following section can proceed (see Section 2.4). If a favorable ratio is not obtained, then this planning process should be terminated.

Secondly, what fraction of the program cost will be directed at evaluating (assessing) the effects of the program?

Some evaluation of the effects of the seeding is strongly encouraged, and very probably essential to the long-term survival of the program (see

Section 2.2 and 2.4). The effort spent on the evaluation will ultimately reflect the level of proof the program sponsors feel they need to realize. Larger programs, especially those conducted with public funds, often require a greater level of proof (more evidence, either physical, statistical, or both) that the program is effective.

6.3.3 Statement of Program Expectations (Likelihood of Success)

Clear statement of the program scope and how the objectives fit within the overall role and goals of the sponsoring entities is essential. For example, a public utility might have as its primary objective, "the comprehensive development of hydropower, maximizing the benefits to the company while minimizing environmental impacts." A weather modification program conducted during the winter months to increase snowpack (and ultimately runoff) may be an alternative that would supply additional streamflow and generating head. A program conducted during the warm season to increase rainfall (and ultimately soil moisture) or to recharge aquifers, may be an alternative that would supply additional plant needs and reduce irrigation requirements, including the mining of groundwater. In many cases, implementation of a cloud seeding program may offer the least costly means to meet such objectives. Before the program is designed, the sponsors should thus state what the program is expected to accomplish and how it will fit into existing water management programs and goals.

Evaluation (assessment) of cloud seeding programs is imperative. A reasonable plan for evaluation should be devised for every program, even if the plan initially only provides for the collection of relevant data. Without meaningful long-term program assessments, cloud seeding programs are invariably discontinued for lack of evidence of effectiveness.

Comprehensive program planning and development need not be the purview of a single agency or entity. In fact, a consortium of groups and various governmental agencies could in many instances carry the development forward faster. This is especially the case if the program sponsors include public entities such as municipalities or counties.

Water development decisions pertaining to cloud seeding must be fact-based. Emotion, suppositions, or political ambition should not govern the final decision (Linsley and Franzini, 1964). The design of most cloud seeding programs requires the accumulation and study of a substantial body of meteorological, hydrologic, economic, environmental, and social data. The services of specialists in each of these fields are needed to collect and interpret these data (Keyes et al., 1995). For most programs, the data are primarily historical, for example, long-term temperature, wind, and precipitation data. The longer the periods of record, the greater is the value of the data.

6.4 PROGRAM DEFINITION

After the completion of the feasibility study or initial program assessment (evaluation), the prospective program sponsor(s) will know if local cloud characteristics and estimated benefit/cost ratios warrant implementation of a precipitation increase cloud seeding program. The scope of the program will be defined by the findings of the feasibility study and should include measures for assessment of the program's results. Potential long-term impacts should also be weighed.

6.4.1 Seeding Modes and Agents (Design)

The design, which is based on the feasibility study, will define the following:

1. *Seeding mode(s):* Plans will incorporate ground-based and/or airborne seeding, as described in Section 5, according to the findings of the feasibility study. Many programs use both, but not necessarily at the same time. For example, a winter orographic seeding program might deploy ground-based seeding devices to target the majority of the supercooled liquid water found at low altitudes over the mountain crests, but might augment this seeding with treatment by aircraft flying upwind of the barrier over a valley, at or near the altitude of the crest line. The aircraft might also be used to target specific convective cloud elements, common in the fall and spring, which may not always be reliably targeted with ground-based facilities. Programs intended to treat convective (warm season) clouds with few exceptions rely on aircraft-based delivery systems.
2. *Siting of equipment:* Ground-based equipment may be sited in accordance with climatological assessments of the prevailing weather parameters important to the transport and diffusion of the seeding material over the intended target area. Validation of the siting could be achieved during program operations through silver and/or trace chemistry analyses as discussed in Section 5.3.1.6. In many locations, particularly in the United States, many of the preferred ground-based generator sites are on public (Forest Service) lands. Access to and permission to use these sites can require a great deal of paperwork, and the final permission (or denial) can sometimes be determined by the personalities of those involved. In general, it is easier to site equipment on private land, when quality, privately owned sites can be found.

   When considering the locations of ground-based seeding facilities and the probable flight tracks of airborne seeding runs during the design phase for winter orographic programs, people are encouraged to use numerical modeling and/or tracers to ensure effective placement, assuming there are adequate financial resources to do so and program goals include that level of effort on the issue. In modeling simulations, it is not sufficient to model only a single wind regime. As many wind profiles associated with precipitation and/or the presence of supercooled clouds should be modeled as possible. This will allow the design of a targeting plan that will maximize the coverage of the target area(s).

3. *Aircraft Base of Operations:* The basing of aircraft is also of some concern, so that the aircraft can safely depart from and return to the airport selected during weather conditions typical of operations. Suitable alternate airports must be available, in the event that the "home" airport is not accessible. In programs conducted in warmer locales or during the warm (convective) season, aircraft that accumulate significant ice can simply descend to warmer levels, shed the ice, and return to operations. However, in many wintertime programs, particularly those in higher latitudes or the interior of continents, this is not possible, for the temperature of the atmosphere all the way to the surface may be less than 0°C. In this case, the option of descending to shed ice does not exist, and the aircraft must simply cease operations once the pilot believes the aircraft has accumulated all the ice it can safely handle. This is true even of aircraft certified for flight in known icing conditions.[15] For programs using multiple aircraft, basing decisions must be considerate of minimizing time from deployment to arrival at the target cloud, as the seeding "window of opportunity" is small in convective clouds. Distribution of aircraft at multiple bases (as available airport facilities allow) is the preferred method in this scenario (Langerud and Gilstad, 2004).
4. *Seeding Agents:* Various cloud seeding agents are discussed in detail in Section 5. The design will identify which agent(s) should be used, and when.

6.4.2 The Evaluation Plan

When a program is defined, an evaluation plan should be part of the program. Most operational cloud seeding programs for the last 30 years or more in warm and cold seasons, especially those in the United States, have decided to seed every cloud or storm meeting their program design criteria as being treatable. This means that with these programs, the only

[15]Certification for flight in known icing does not mean that an aircraft can carry an infinite amount of ice, nor that the aircraft can rid itself of ice sufficiently to remain in icing conditions indefinitely. Rather, certification for "known icing" means that the aircraft has demonstrated the ability to safely fly within cloud condition known to produce icing long enough to safely pass through the icing level.

avenue left for evaluation has been to compare the precipitation received within the target area to that received in a nearby, climatologically similar "control" area. If a long-term positive correlation in precipitation between the target and control areas can be established for those years when no seeding was done in either, then a subsequent relative change in the target can be attributed to seeding. The risk in drawing such a conclusion is that climates change, though changes in the relationship between nearby areas may be less than changes in the overall precipitation climate, or one area or the other may have been affected by an extreme event that didn't affect the other. Such things could lead to the wrong conclusion—either positive or negative—being drawn about the effects of seeding.

A second option, seldom used in operational programs, is to randomize treatment for the intended area of effect. That is, to decide randomly which days or storms should be treated, and to compare them to those that were not. This is usually the approach adopted in scientific experiments, but this course has its drawbacks as well. While it can eliminate the possibility that climatological changes were the reason for any observed changes in precipitation, the chance of a single extreme event (seeded or not seeded) having a dominant effect on the conclusion still exists. Another drawback is that in order to randomize, some fraction of all suitable clouds must not be treated. This, of course, diminishes the overall effects and thus the benefit/cost ratio of the program.

Randomization of treatment for a few seasons as a program is first started should perhaps be considered if the sponsor is sincerely interested in establishing the strongest possible evidence of program effectiveness (Mooney and Lunn, 1969). Historically, most randomized programs have been conducted as research programs, but there is nothing that precludes a serious operational program from randomizing treatment if its sponsors choose to do so. Randomization need not be 50:50, that is, one case seeded for every case not seeded, but could be 2:1, wherein two out of every three cases is still seeded. This lengthens the period needed to draw firm conclusions, but allows most seedable cases to be treated, increasing the program's immediate impact.

Barring this, one is usually compelled to examine observable differences between those clouds or systems that are seeded and other clouds or systems, presumably of similar nature and potential, which are not seeded. Problems often arise because people involved in seeding programs naturally seed those situations perceived to have the greatest potential to produce precipitation, and other biases sometimes known and sometimes unrecognized, exist (Woodley and Rosenfeld, 2004).

The initial step in program evaluation is the collection of relevant precipitation data, and whatever supporting atmospheric data are obtainable. Some scientists believe that proper evaluation of a weather modification program can only occur if the program is randomized (e.g.,

National Research Council, 2003). Others hold that meaningful results can be obtained for properly designed non-randomized programs, especially if such programs are coupled with physical measurements that verify some of the key processes involved, such as those related to targeting and nucleation (Orville et al., 2004). Such physical data may include cloud physics measurements, radar and satellite data, numerical modeling to aid targeting, and trace chemistry measurements (e.g., Changnon et al., 1979). Some of these techniques may be more readily applied to research rather than operational programs.

Some newer approaches thus combine the target vs. control approach with physical measurements, computer modeling, and/or trace chemistry to validate program operations and verify that the clouds intended to be seeded were suitable candidates and were actually seeded. For example, when the project finds silver concentrations in snow within the target that fell during a period of seeding, the argument that seeding increased the snowfall is strengthened. This alone does not constitute unequivocal proof; it is possible that the silver was "scavenged" by falling snowflakes, and was not the origin of the ice crystals that formed the snow. The tracer techniques discussed in Section 5 can resolve that question.

New, higher-end computer models such as MM5 and the WRF may allow predictions of where seeding agent released from specific locations should travel and be deposited. Thus, the use of a model like MM5 on even a non-randomized seeding program could allow improved definition of both target and control areas, and allow for storm-by-storm evaluations to be made. For example, the model might identify situations in which the control area was contaminated by agent drifting out of the target region, or in which the target area was not well-targeted. Before MM5, WRF, or any model should be relied upon for such applications, some validation should be performed for the region of interest. For example, this could be done by checking for silver or other tracers in new "seeded" snow where the model says it should be found. Without adequate model validation, the comparatively small anticipated differences in precipitation due to seeding will fall within the level of uncertainty associated with the modeling guidance, resulting in inconclusive results.

Correlations between target and control areas are normally weaker the shorter the sampling interval. In other words, correlations for individual storm periods are weaker than those for individual months, which are weaker than those for an operational season of several months or longer. The degree of correlation determines the ability of a target/control evaluation method to discriminate the seeding signal from the natural variability of precipitation measurements (i.e., a signal to noise problem). If affordable, even on a temporary basis, modeling is a tool that can reduce the "noise" in the signal by clarifying which areas were actually targeted during a given storm event.

In some long-term water development programs, it is not uncommon to have some portions of the program go essentially unchanged for 50 to 100 years (Kulper, 1971). Water managers should periodically review all the objectives and how they are being addressed, so that long-term program components are updated as knowledge and technology allow. This process will allow water managers to consider weather modification if/when appropriate circumstances allow.

In many long-term cloud seeding programs, seeding may have been conducted in various ways for 10 to 20 years, and the data collected then studied to evaluate the program effectiveness. Greater consistency of the operational methodology would reduce the "noise level" in evaluation, though few would argue that a desire for consistency should impede the adoption of improved technology in an operational program. In many cases it has taken years to analyze the different methodologies, with much debate about the findings. Often, no final conclusion has been reached regarding program efficacy.

It is very important that the parameters upon which evaluation (assessment) is to be based be defined before the program begins, and any necessary equipment (e.g., precipitation gauges) be deployed. The addition of measurements like those obtained from precipitation gauges will only be useful in a systematic program evaluation if, for example, a similar number of gauges are also deployed within the proposed control area and the project is randomized. Without randomization, there will likely be no unseeded data available for comparison with the seeded data from the newly installed gauges. In such a case, siting criteria (elevation, exposure) must be the same for all supplemental gauges.

In addition to examination of precipitation data, other techniques are also available. Any physical measurements made during both natural and seeded storms help document the development of precipitation. Sampling of snow in and around the target area for evidence of seeding agent(s) can help verify correct targeting, further strengthening results, although this expense is not trivial.

Radar reflectivity data can be used to estimate precipitation, especially during warm-season programs when most of the precipitation is in liquid form at lower levels. Such estimates are derived from empirical relationships between radar reflectivity, Z, and rainfall rate, R, termed Z-R relationships. These relationships can be "tuned" for a best fit to each local area, or even on a storm-by-storm basis. This is accomplished by comparing the radar-estimated precipitation with that actually observed by surface gauges, and then "adjusting" the radar estimates to make them agree with the surface observations. The Z-R relationships should generally hold for those storms that form under similar conditions to those present when the relationships were developed. Z-R relationships that yield poor estimates should be reported to the program manager, who should inform the scientists so that they can develop better the relationships for future operational use in the particular area. Once the Z-R relationships are accomplished for a particular storm, the radar can be used to estimate precipitation, providing estimates of total rainfall amount over the area covered by the radar (e.g., Woodley et al., 2001). Such area-wide estimates are difficult to make solely with rain gauges, because they are normally many kilometers (miles) or tens of kilometers (miles) apart.

A new radar-based technique that utilizes the advantages of the WSR-88D 10-cm (S-band) wavelength NEXRAD weather radars operated by the National Weather Service in the United States has been described by Woodley and Rosenfeld (2004). This technique uses "floating" target units that are tracked by the radars.

Because weather patterns vary considerably from season-to-season, and invariably program sponsors and operators learn how to better conduct various program facets, operational and evaluation plans should be flexible and seasonally subjected to review and modification if appropriate. During the initial season or two, procedural modifications may be made even more frequently. The risks associated with making programmatic modifications should be estimated and passed on to the program sponsors prior to their implementation. Usually, the greatest complication arises from the need to evaluate operations differently if they are conducted differently. Early on, this may not be a major issue, but if significant changes are made a year or more into the program, it may be necessary to reevaluate the program to ensure that apples are not compared with oranges.

Evaluation of weather modification programs has proved to be difficult. The primary difficulty arises from the unavoidable fact that no two clouds or even two storm systems are exactly the same, and one cannot simply treat (seed) one, and not the other, and then observe the differences. Evaluations on a cloud or storm basis are now possible with the use of analytical numerical cloud models that can provide forecasts (or even hindcasts) of specified response variables. Evaluation statistics comprising differences between observed and forecast variables should be less noisy than the observed variables themselves. Such models are becoming increasingly helpful; but the answers are only as good as the model(s). As previously stated, program-specific validation/verification of the model should be performed. For examples, see Orville et al., 1984; Helsdon and Farley, 1987; Rasmussen and Heymsfield, 1987; Farley et al., 1989; Kopp et al., 1990; Orville and Kopp, 1990; Kopp and Orville 1994; Stith et al., 1994; and Wilhelmson and Wicker, 2001.

Sponsors of operational programs often do not expect to conduct evaluations that they would consider necessary by the scientific community to "prove" that cloud seeding is working as intended. A quote from a paper by Bruintjes (1999) illustrates this point: "The fact that many operational programs have been going on and have increased in number in the past 10 years indicates the ever-increasing need for additional water resources in many parts of the world, including the United States. It also suggests that the level of proof needed by users, water managers, engineers, and operators for the application of this technology is generally lower than what is expected in the scientific community. The decision of whether to implement or continue an operational program becomes a matter of risk management and raises the question of what constitutes a successful precipitation enhancement program. This question may be answered differently by scientists, water managers, or economists depending on who answers the question. This difference is illustrated by the fact that although scientific cloud seeding experiments have shown mixed results based on the level of proof required by the scientific community, many operational cloud seeding programs are still ongoing. However, it also emphasizes that the potential technology of precipitation enhancement is closely linked to water resources management. It is thus important that the users of this potential technology are integrated into programs at a very early stage in order to establish the requirements and economic viability of any program. In addition, the continued need for additional water and the fact that most programs currently ongoing in the United States and the rest of the world are operational programs emphasizes the need for continued and more intensive scientific studies to further develop the scientific basis for this technology." Another quote from Silverman (1978) supports the quote from Bruintjes: "Users of weather modification are shrewd business people. They understand that they are, in many cases, taking a gamble when they use weather modification, but it is no greater risk than they take in other aspects of their business."

The Weather Modification Association's Code of Ethics (Weather Modification Association, 2004) states: "Evaluations of programs are strongly encouraged. Any limitations to evaluation will be reported to the client. Procedures to be used in evaluations should be specified in advance."

6.4.3 Quantification of Findings

It is important to assess the effectiveness of cloud seeding for precipitation augmentation. Evaluation (assessment, at times estimation) of the effects of seeding efforts might seem at first thought to be a relatively straightforward exercise. However, the more commonly stated percentage increases, unfortunately, fall well within the normal range of natural precipitation variability. Thus, quantifying the differences attributable to cloud seeding becomes a challenging enterprise. The difficulty in doing so does not mean that evaluations should not be done. Rather, it indicates that the whole matter should be addressed from a perspective that reflects an understanding of (1) the evaluation possibilities (statistical and physical) and their limitations; (2) some primary pitfalls involved; (3) the costs of the various methods and what they can realistically be expected to provide; and (4) a balanced plan for evaluation that fits the program's goals and needs. This is an issue worthy of careful consideration. At the heart of the matter is the "level of proof" issue and fundamental benefit/cost considerations.

As previously noted, most operational programs are not randomized. Thus, after the completion of a month's or a season's seeding activities, the precipitation recorded within the target area(s) often is compared with that observed in the control area(s). Judging program outcome by comparing data from just a single month or season is risky—as natural variability can be considerable, and it is possible that one or two large events over either the target or control area(s) could greatly affect the perceived program impact. For example, a large event over the control area(s) when no event of similar scale occurred over the target(s) could wrongly lead one to conclude that the effect of the program was negative. Conversely, a large event over the target(s) when none occurred over the control(s) could incorrectly lead one to conclude that an exaggerated positive effect resulted from seeding. In either case, the apparent conclusion is inaccurate, simply the result of a "bad draw" due to naturally large events that don't occur over both target(s) and control(s).

For example, consider a program wherein mountain storms moving from west to east are usually seeded. Suppose a deep low pressure center passes to the south of the target, creating much vertical motion, condensation, and precipitation over the target area, along with easterly winds. If the program is set up to seed storms passing from west to east, seeding a storm characterized by easterly flow certainly will have little impact in the target area(s). If precipitation is considered on a storm-by-storm basis, such storms can be identified and excluded from the analysis. If it is not, then the resulting precipitation would be included—even if most of it fell outside the target area(s), perhaps even over the control(s).

Because of the potential for suggestion of bias or conflict of interest in evaluations, statistical and/or physical evaluations of operational programs could be conducted by qualified but disinterested third parties, i.e., people with no stake in the outcome. Such people are more readily able to identify possible sources of bias, and are more likely to offer criticism when criticism (often constructive) is warranted. The independence associated with third-party evaluation may strengthen program credibility. The sponsors and public may be more accepting of a third-party report than an in-house document, regardless of the degree of honesty and expertise with which the in-house report was assembled. The WMA Statement on Standards and Ethics (loc. cit.) encourages program evaluations. It should be recognized and emphasized that if a third-party evaluation of a non-randomized program is to be undertaken, the evaluation methods and procedures should be established in advance (a priori) of the seeding period to be evaluated. For example, if a target and control winter analysis is proposed, the stations making up the target and controls and the historical period to be used should be specified in advance. This is important because otherwise there can always be the question of bias in the evaluation of a non-randomized program if the basis for the evaluation is established only after the fact (a posteriori).

Care must be taken in the selection of the party doing the evaluation, to ensure that the party is (1) qualified and (2) truly disinterested in the result. The Weather Modification Association maintains a list of corporate members, many of whom are very capable of such analyses.

Collectively, there are many ways to assess operational weather modification programs, even without randomization. The critical issues are to incorporate evaluation methods that are appropriate to a program's needs and goals and to consider carefully the "level of proof" that is required for program support decision making. One must remain mindful of the fact that every evaluation method known has some limitation(s) that can be noted. Thus, absolutely conclusive quantitative proof of cloud seeding effects is not currently achievable in the realm of non-randomized operational programs. Recognizing that, at this time in the evolution of cloud seeding technology, it is not possible to deal in absolutes regarding the evaluation question, a program manager can survey the spectrum of assessment possibilities and, perhaps with the help of an expert weather modification consultant, select an evaluation method that best fits program needs. For new programs, it is advantageous to establish and state the method(s) before seeding begins and, as stated earlier, an independent evaluation can be considered as a primary or secondary tool. One can thus help to avoid obvious biases and bolster program credibility. New technologies for assessment can be considered and incorporated if deemed appropriate and cost effective.

6.5 PROGRAM CONTROL

The infrastructure of an operational cloud modification program should be clearly set forth in a program operations plan, written well in advance of the onset of seeding operations and reviewed by the sponsor, contractor, and any regulatory agencies having jurisdiction over weather modification. Such operations manuals describe the communications "chain of command," operations procedures, seeding suspension criteria, and reporting requirements along with operational theory and the theoretical (scientific) basis for the program. In addition, a summary of the program's environmental review is often contained; if no program-specific review was conducted, an overview of the environmental findings from similar programs may be provided. The operations plan thus is a one-stop source of information about the program, containing most program-related information, apart from results, which will be determined later.

6.5.1 Seeding Decisions

Operational weather modification programs are controlled by guidelines and other criteria established by the sponsor, and conveyed to the program operator (contractor), usually in the person of an operations director or program manager. The general flow of information, as described in ASCE (2003), is illustrated in Figure 6-2.

The daily flow of program decision making is generally focused at an operations center. For warm season (summertime) programs, this is often also the program radar site. For cold season programs it may be a room dedicated for the purpose within the sponsor's offices, a remote field location where other program equipment is sited (a microwave radiometer, for example), or, if aircraft are involved, an office at an airport.

In Figure 6-2, the core group involved in the actual day-to-day conduct of operations includes those at the operations center, the operations personnel, and those responsible for the maintenance and servicing of program facilities. The people in these three groups typically communicate at

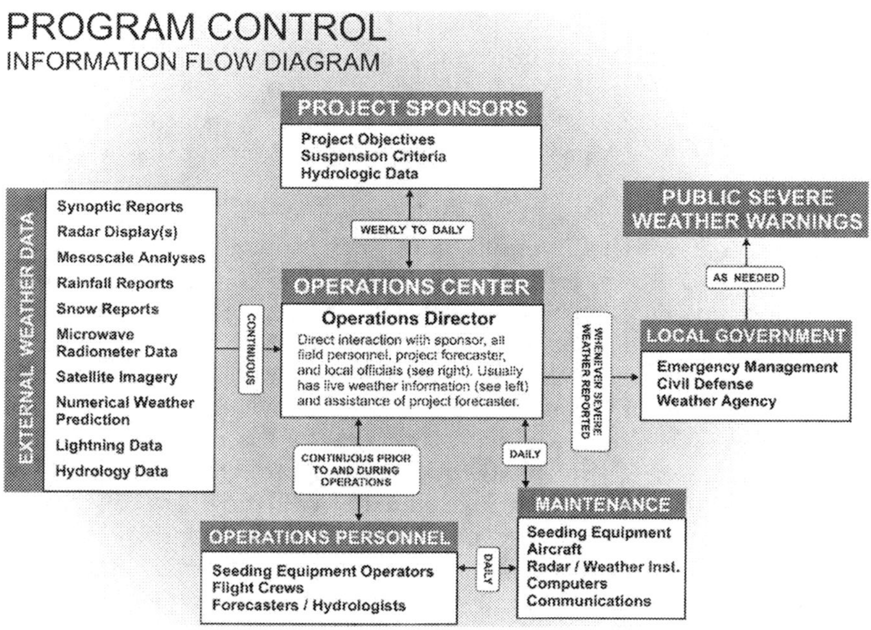

FIGURE 6-2. Communications Infrastructure of a Typical Weather Modification Program. Some items shown are season-dependent and were revised as shown in ASCE (2003).

least daily, and often many times a day when the program has the potential for seeding activities. Maintenance and servicing of field sites is usually conducted when inclement weather is not occurring. Therefore, the maintenance personnel are generally most active when the operations personnel are not, and vice versa.

Additional input is received frequently at the operations center from the program sponsors. Often, this communication may relate to updated hydrological data (reservoir levels, streamflows, etc.) or updated rainfall or snowfall reports, including snow surveys. The program sponsor always has the prerogative to suspend operations, in addition to the program seeding suspension criteria that are pre-established in the program design. It is also prudent to grant the seeding operations director unilateral real-time suspension authority as a safeguard against rapidly developing hazardous conditions when communication among the usual parties may not be readily accomplished. An example of the value of this arrangement is shown later in this section.

6.5.2 Data Collection and Access

Data collection is a critical part of any evaluation program, as discussed in Section 6.4. The data collected and archived are invariably a function of the design of the evaluation plan, restrictions on the data collection (frequency of snow sampling, for example), and the funding level of the program. Additional instrumentation and data collection over and above what was done prior to the onset of the program may help evaluate the program efficacy, and ultimately, the economics of the program, but often are not affordable. Section 5.0 contains additional discussions on this topic.

It can be helpful to establish a public access web site offering access to real-time program data. Typically, such sites include real-time program radar imagery, status of seeding (aircraft or ground-based), seasonal seeding and precipitation statistics, the daily program forecast, frequently asked questions and answers, and an e-mail address to which inquiries and comments can be sent. Such web sites provide the interested people with real-time information, imparting real knowledge of what is going on. This reduces speculation, and eliminates any sense of "secrecy" or covertness about the program. If a frequently updated program site is not considered practical by sponsors, they may find another way to provide regular, timely program information if appropriate to program goals and needs.

Some programs have ended because of lack of evaluation and/or lack of a rapid communication during potentially hazardous conditions. The program protocol should have clear operational criteria and restrictions related to short-term and long-term suspension of the programs, and equally clear evaluation objectives.

6.5.3 Seeding Suspension Criteria

Sections 2.2.4 and 3.2.3 of this manual both point to the need for programs to implement safeguards that will ensure the public safety and environmental well being. The most common such safeguard is a well-designed set of criteria that, when satisfied, triggers an immediate cessation of seeding activities. Such safeguards differ for warm season (convective) and cold season (orographic) seeding programs, but share the same common goals.

6.5.3.1 Warm Season Suspension Criteria. Warm season programs designed to increase precipitation deal most commonly with convective storms. Such storms range in scale from cumulus congestus (towering cumulus) clouds that produce little, if any, precipitation that reaches the ground to heavy thunderstorms capable of producing extreme precipitation volumes in very short times. Rainfall rates from deep moist convection commonly exceed 25 mm (1 inch) per hour (Lamb, 2001), but sometimes are far greater. For example, 305 mm (12 inches) of rain fell in Holt, Missouri, on June 22, 1947 in just 0.7 hour (Lott, 1954). The hazardous implications for such heavy rainfalls are further complicated by the fact that, oftentimes, such storms are slow moving and release much of their precipitation over the same area, as in events that result in extreme natural floods that occurred in Rapid City, South Dakota (Maddox et al., 1978; Doswell et al., 1996), and the Big Thompson Canyon of Colorado (Maddox et al., 1978). Such storms/convective systems ought not to be seeded. Thus, suspension criteria need to provide for the early identification of storms with such potential, so that seeding may be called off in advance of the event.

Some factors that alert program forecasters to the potential for warm season flooding by convective storms are listed below. The likelihood of potential flooding increases as more of these factors are present. They include:

1. Convective instability as indicated by upper-air soundings (large convective available potential energy)
2. Atmospheric dynamics that favor storm development
3. Precipitable atmospheric water in excess of 25 mm (1 inch)
4. Low level wind shear profiles that favor moisture advection in the low levels, and "ventilation" of the storms aloft
5. Light to moderate mid-level winds (will favor slow storm movement)
6. Saturated or near-saturated soils
7. Reservoirs or other water control structures that are at or near storage capacity

Together, these and other local factors can alert the program forecaster to the increased probability of localized flooding and the potential need to suspend seeding. A rule of thumb often used by program managers is that if the forecaster sees enough signs to begin to worry about localized flooding, there is sufficient cause to suspend seeding, even in the absence of any storms. Such decisions are usually made by the program manager (operations director) after consultation with the program forecaster and the program sponsor(s). A good example of pre-storm suspension of warm season seeding occurred in June of 2002 in a Texas precipitation enhancement program (Resler et al., 2002) when the program operations director (and also forecaster) became concerned early in the day about flooding potential. This particular event, in Abilene, occurred on a weekend, and the program sponsor could not be contacted. The operations director consulted with his immediate supervisor within the company, and the decision was made to ground all seeding aircraft, thus suspending all seeding even before a cloud formed. Later that afternoon and evening, numerous slow-moving, heavy precipitation-producing thunderstorms developed that produced the flooding of numerous streets and basements in portions of Abilene, but no seeding had been done. In this case, the program sponsor was not contacted before the decision to suspend was made, as would normally be the case. However, the "when in doubt, suspend" rule of thumb was appropriately applied.

There are additional real-time indicators that warm season seeding operations to increase rainfall should be suspended. These include:

1. Flood watches or warnings issued by local authorities, e.g., the National Weather Service.
2. Radar indications of slow-moving or stationary thunderstorms that are producing heavy rains.
3. Radar-derived precipitation estimates that indicate excessive rainfall, particularly if soils are known to be saturated or near saturation.
4. Train echoing, when a series of storms follow the same path, one after another. Though no single storm may produce enough rain to result in flooding, the collective effect of a series of storms often does. Train echoing (or training) often occurs along stalled or slow-moving surface frontal boundaries.
5. Tornadic storms are not seeded in some programs, but are in others. One school of thought holds that one should do nothing (i.e., seeding) that could be perceived as contributing to a dangerous situation. Another is that because there is no direct evidence in the literature that indicates seeding can cause or exacerbate tornadic storms, seeding should continue.

The seeding suspension criteria developed for any warm season program should address all of these factors, and should clearly state the threshold for suspension. Some suspensions (as noted above) occur before any storms even develop, while others occur when storms having flooding potential are actually observed. In either event, thorough records should be kept of all suspensions, including the time they were ordered and the reasons why. Some programs have determined that local news media should be notified when suspensions occur. Others have even sought the visual verification of suspension of operations by local authorities, e.g., local law enforcement has been asked to visit an airport to verify the aircraft are not flying. The latter is not generally required unless there are known to be people in the program area who would seek to blame any natural misfortune on the seeding program—in which case verification of inactivity by law enforcement or other civil authority will quickly dispel any such claims.

6.5.3.2 Cold Season Suspension Criteria. Cold season programs designed to increase precipitation deal most commonly with orographically induced, often synoptically aided clouds and storms. Such storms range in scale from simple "cap clouds" that barely enshroud the mountain tops, to energetic synoptic-scale winter storms capable of producing extreme precipitation volumes (rainfalls or snowfalls) in relatively short times.

Suspension criteria are also needed for cold season programs to avoid even the appearance of exacerbating severe winter conditions. Common negative consequences of better-organized cold season storms include: snowy, icy, and/or blocked roadways, reduced visibility, heightened avalanche potential, increased roof loading, and additional snow-removal costs. All of these effects can also occur with large natural storms. Some winter programs are conducted to increase rainfall, not snowfall. As a consequence, some of the same type of suspension criteria developed in summer programs may be needed in these operations.

The cold season suspension criteria are intended to avoid:

1. The appearance of contributing to extreme precipitation events, which are usually naturally efficient. To do this, one must identify potential extreme snow and/or severe winter weather events prior to their development over the target, so that they are not seeded.
2. Some storms are cold enough to be naturally efficient precipitation producers anyway, and seeding may result in little additional precipitation, as suggested by Boe and Super (1986) and Super and Boe (1988); therefore many cloud storms are not seeded.
3. Contributing to already heavy snowpack, such that spring runoff may become excessive and/or unmanageable.
4. Seeding very warm storms that may produce heavy rains on existing snowpacks at high elevations.

Suspension for naturally heavy storms is generally based on the program forecaster's experience with the local climatology, prognostic charts (especially numerical models), and real-time observations of developing storm conditions. The last of these is greatly aided by real-time satellite imagery.

Suspension to avoid contributing to excessive snowpack is more readily accomplished. Available snow course data are periodically compared to "normal" snowpack to determine the percentage of normal snowpack present. Program-specific criteria should be established (and set forth in the operations plan) that specify when seeding is to be suspended. For example, early in the snow season (say, December), the total snowpack (in a northern hemisphere program) is usually a small fraction of the annual total. Therefore, seeding may not be suspended until measured snowpack exceeds perhaps 200% or 250% of "normal" for that date. As the season progresses and the total snowpack increases, the suspension threshold may then drop, such that by the end of January the suspension threshold might be 175% of normal, and 150% of normal by the end of February. By the first of April, seeding may be suspended if snowpack exceeds 125% of normal. These specific numbers are provided simply as examples, and area-specific appropriate values should be determined by hydrological studies of each local area. Suspension of seeding because snowpack meets or exceeds a certain percentage of normal does not necessarily mean that the program is shut down for the balance of the season, only that seeding operations should cease until such later time as the percentage of normal drops below the given threshold for the date. If the weather pattern remains wet for the duration of the winter, seeding may not resume, but if drier conditions persist for a time, as is often the case, seeding then resumes.

6.6 PROGRAM MANAGEMENT

The program management has certain ongoing responsibilities, some of which continue beyond the operational season. While some program sponsors prefer that the people or entity (contractor) conducting the actual seeding make most day-to-day decisions with minimal oversight, others are more hands-on and desire more direct involvement in day-to-day decision making. Such determinations are made by the program sponsor. Invariably, it is the responsibility of the program sponsor to ensure the following:

1. *Oversight of operational decision making:* This need not be micromanagement, but should include a mechanism through which operational decisions are periodically reviewed.

2. *Recordkeeping:* In most cases the sponsor requires that detailed records be kept of operations. In many cases, such records are required by law. For example, in the United States, the National Oceanic & Atmospheric Administration requires that the type, location, and duration of operations be periodically reported, along with quantities of seeding agent dispensed.
3. *Public information and outreach:* Program sponsors often have public relations staff available that can include information about cloud seeding efforts and effects in their presentations, or who at least can be prepared to answer questions. Many sponsors also include the expertise of the contractor in this effort. This might entail public talks, management of a project web site (see Section 6.5.2), or the preparation of non-technical program informational brochures.
4. *General assessment (evaluation) of the program:* While some sponsors are satisfied with program evaluations conducted by the contractor, others choose to retain expertise from outside the program to evaluate program efficacy. The reasons for the latter are twofold. First, even if an "in-house" evaluation is done with utmost care and effort to avoid biases, it is very common for such evaluations to be called into question simply because they were conducted by the same party doing the seeding. Some dismiss in-house evaluations as being self-serving. The second reason is that evaluation by a qualified individual or team that has no direct stake in the outcome is less likely to be biased (consciously or subconsciously) and more likely to be accepted by the public. In either case, the key is that the person(s) doing the evaluation be qualified and unbiased.
5. *Disclosure of findings:* Once evaluation is completed, findings should be released to the press and made readily available to the interested public. Evaluations are often posted on the sponsor's or contractor's web sites. Evaluation based upon a single year's efforts will in most cases be very difficult, as the magnitude of the anticipated effect is within that of natural variability. This is most certainly true of non-randomized programs. Therefore, disclosures concerning operations after the first few seasons may be limited to a summary of operational days and any physical measurements made that reflect upon the program efficacy.
6. *Modification of operational design:* If difficulties are discovered concerning the design of the program, it is within the sponsor's purview to ensure that the difficulties are corrected. If seeding criteria or methodology are modified it might complicate comparisons of findings from the initial season(s) with those of future season(s), but it is more important to correct problems.

The importance of open program information dissemination cannot be overstated. Beyond a program web site as described in Section 6.5.2, weekly and/or monthly program updates and educational news releases for decision makers such as county commissioners, legislators, governors, and administrative staffs can be used to keep the program visible and the public informed.

6.7 REFERENCES

American Society of Civil Engineers (2003). *Standard Practice for the Design and Operation of Hail Suppression Projects*, ASCE/EWRI Standard 39-03, ASCE, Reston, VA, p. 40.

Boe, B.A., and Super, A.B. (1986). "Wintertime characteristics of supercooled liquid water over the Grand Mesa of western Colorado." *J. Weather Modification*, **18**, WMA, Fresno, CA, 102–107.

Bruintjes, R.T. (1999). "A Review of Cloud Seeding Experiments to Enhance Precipitation and Some New Prospects." *Bull. Amer. Meteor. Soc.*, **80**, No. 5, AMS, Boston, MA, 805–820.

Changnon, S.A., Huff, F.A., and Hsu, C.F. (1979). "On the Need to Evaluate Operational Weather Modification Projects." *Bull. Amer. Meteor. Soc.*, **60**, (7), AMS, Boston, MA, 770–774.

Doswell, III, C.A., Brooks, H.E., and Maddox, R.E. (1996). "Flash flood forecasting: An ingredients-based methodology." *Weather Forecasting*, **11**, 560–581.

Farley, R.D., Price, P.E., Orville, H.D., and Hirsch, J.H. (1989). "On the Numerical Simulation of Graupel/Hail Initiation via the Riming of Snow in Bulk Water Microphysical Cloud Models." *J. Appl. Meteorology*, **28**, 1128–1131.

Griffith, D.A., and Solak, M.E. (2002). "Economic Feasibility Assessment of Winter Cloud Seeding in the Boise River Drainage, Idaho." *J. Weather Modification*, **34**, WMA, Fresno, CA, 39–46.

Heimbach, J.A., and Super, A.B. (1988). "The Bridger Range, Montana, 1986-1987 snowpack augmentation program." *J. Weather Modification*, **20**, WMA, Fresno, CA, 19–26.

Helsdon, Jr., J.H., and Farley, R.D. (1987). "A Numerical Modeling Study of a Montana Thunderstorm: Part 2. Model Results versus Observations Involving Electrical Aspects." *J. Geophysical Research*, **92**, 5661–5675.

Keyes, Jr., C.G., Warburton, J.A., and Renick, J.H. (1995). "How to Implement a Cloud Seeding Program," Section 6, *Guidelines for Cloud Seeding to Augment Precipitation*, Manual of Professional Practice No. 81, Co-editors Kahan, et al., ASCE, Reston, VA, 125–136.

Kopp, F.J., Orville, H.D., Jung, J.A., and McNider, R.T. (1990). A Parameterization of Radiation Heating at the Surface in a Numerical Cloud Model, *Preprints, 1990 Conference On Cloud Physics*, San Francisco, AMS, Boston, MA, J81–J84.

Kopp, F.J., and Orville, H.D. (1994). The Use of a Two-Dimensional, Time-Dependent Cloud Model to Predict Convective and Stratiform Clouds and Precipitation, *Weather and Forecasting*, **9**, 62–67.

Kulper, E. (1971). *Water Resources Project Economics*. Butterworth and Company, London, England.

Lamb, D. (2001). Rain production in convective storms. American Meteorological Society Monograph, **28**, No. 50, *Severe Convective Storms*, AMS, Boston, MA, 299–321.

Langerud, D.W., and Gilstad, A. (2004). *North Dakota Cloud Modification Project Operations Manual*, North Dakota Atmospheric Resource Board, Bismarck, ND, 56 pp.

Linsley, R.K., and Franzini, J.B. (1964). *Water-Resources Engineering*. McGraw-Hill Book Company, New York, NY.

Lott, G.A. (1954). The world-record 42-minute Holt, Missouri, rainstorm. *Mon. Wea. Rev.*, **82**, 50–59.

Maddox, R.A., Hoxit, L.R., Chappell, C.F., and Caracena, F. (1978). Comparison of meteorological aspects of the Big Thompson and Rapid City flash floods. *Mon. Wea. Rev.* **106**, 375–389.

Mielke, P.W., Brier, G.W., Grant, L.O., Mulvey, G.J., and Rosenweig, P.N. (1981). "A Statistical Reanalysis of the Replicated Climax I and II Wintertime Orographic Cloud Seeding Experiments." *J. Appl. Meteorology*, **20**, AMS, Boston, MA, 643–659.

Mooney, M.L., and Lunn, G.W. (1969). "The area of maximum effect from the Lake Almanor randomized cloud seeding experiment." *J. Appl. Meteorology*, **14**, No. 1, AMS, Boston, MA, 11–17.

National Research Council (2003). *Critical Issues in Weather Modification Research*. Committee on the Status and Future Directions in U.S. Weather Modification Research and Operations, Board on Atmospheric Sciences and Climate, Division of Earth and Life Sciences. National Academies Press, Washington, D.C., 131 pp.

Orville, H.D., Farley, R.D., and Hirsch, J.H. (1984). "Some Surprising Results from Simulated Seeding of Stratiform-Type Clouds." *J. Climatology Appl. Meteorology*, **23**, AMS, Boston, MA, 1585–1600.

Orville, H.D., and Kopp, F.J. (1990). "A Numerical Simulation of the 22 July 1979 HIPLEX-1 Case." *J. Appl. Meteorology*, **29**, AMS, Boston, MA, 539–550.

Orville, H.D., Boe, B.A., Bomar, G.W., Cotton, W.R., Marler, B.L., and Warburton, J.A. (2004). A Response By the Weather Modification Association to the National Research Council's Report Entitled "Critical Issues in Weather Modification Research," H.D. Orville, Ed. WMA, Fresno, CA, www.weathermodification.org. 54 pp.

Rasmussen, R.M., and Heymsfield, A.J. (1987). "Melting and Shedding of Graupel and Hail. Part I: Model Physics." *J. Atmospheric Sciences*, **44**, AMS, Boston, MA, 2754–2763.

Resler, R.L., Hahn, C.A., and Boe, B.A. (2002). West Central Texas Rainfall Enhancement Program, 2002 Final Report by Weather Modification, Inc., to the West Central Texas Council of Governments, 80 pp + appendices.

Silverman, B.A. (1978). "What Do We Need in Weather Modification?" *J. Appl. Meteorology*, **17**, AMS, Boston, MA, 867–871.

Stauffer, N.E., Jr. (2001). "Cloud Seeding–The Utah Experience." *J. Weather Modification*, **33**, WMA, Fresno, CA, 63–69.

Stith, J.L., Burrows, D.A., and DeMott, P.J. (1994). "Initiation of Ice: Comparison of Numerical Model Results with Observations of Ice Development in a Cumulus Cloud." *Atmos. Res.*, **32**, 13–30.

Super, A.B. (1986). "Further exploratory analysis of the Bridger Range winter cloud seeding experiment." *J. Climate Appl. Meteorology*, **25,** 1926–1933.

Super, A.B., and Boe, B.A. (1988). "Microphysical effects of wintertime cloud seeding with silver iodide over the Rocky Mountains, Part III: Observations over the Grand Mesa, Colorado." *J. Appl. Meteorology*, **27**, AMS, Boston, MA, 1166–1182.

Super, A.B., and Heimbach, A.J. (1983). "Evaluation of the Bridger Range Winter Cloud Seeding Experiment Using Control Gages." *J. Climate Appl. Meteorology*, **22,** AMS, Boston, MA, 1989–2011.

Weather Modification Association (2004). "Statement on Standards and Ethics for Weather Modification Operators." *J. Weather Modification*, **36**, Weather Modification Association, Fresno, CA, 98–104.

Wilhelmson, R.B., and Wicker, L.J. (2001). "Numerical modeling of severe local storms." *Amer. Meteor. Soc. Monogr.*, **28**, No. 50, AMS, Boston, MA, 123–166.

Weather Modification Advisory Board (1978). "The Management of Weather Resources, Volume I, Proposals for a National Policy and Program." Report to the Secretary of Commerce, Washington, DC.

Woodley, W.L, Drori, R., Rosenfeld, D., Orr, S., and Bomar, G. (2001). "Results of monthly and seasonal gauge vs. radar rainfall comparisons in the Texas Panhandle." *J. Weather Modification*, **33**, WMA, Fresno, CA, 46–62.

Woodley, W.L., and Rosenfeld, D. (2004). "The Development and Testing of a New Method to Evaluate the Operational Cloud-Seeding Programs in Texas." *J. Appl. Meteorology*, **43**, AMS, Boston, MA, 249–263.

GLOSSARY

Many definitions are from the Glossary of Meteorology (AMS 2000; 2nd edition, T.S. Glickman ed., AMS, Boston, MA, 855pp), where applicable. Alternative glossary entries are indicated in *italics*. Some definitions are similar in the ASCE Manual No. 81 (Kahan et al., 1995), ASCE (2004), and ASCE (2005) that were published by the ASCE/EWRI Atmospheric Water Management Standards Committee (http://www.ewrinstitute.org) or in other weather modification committees of ASCE.

2D-TD: Two dimensional–time dependent.

Abnormally dangerous activity: Activity involving a high degree of risk which is not a matter of common usage and for which there is legal liability for resulting harm, even though utmost care has been taken to avoid such harm.

Advection: The process of transport of an atmospheric property (i.e., air temperature, water vapor content, or moisture) caused solely by the velocity field of the atmosphere. This movement is usually considered to be in the horizontal direction but may be used in terms of vertical movement. Vertical advection implies motions are predominately vertical and driven by buoyancy forces. It may operate in conjunction with radiation, and it may lead to the formation of clouds or possibly enhance the precipitation process under the right atmospheric conditions.

Advertent weather modification: Weather modification resulting from intentional efforts by humans to change the weather.

AES: Atmospheric and Environmental Service.

AgI See *silver iodide.*

Air mass: A large body of air (hundreds to thousands of square kilometers) that possesses similar temperature and dew point characteristics as a function of height.

AMS: American Meteorological Society, 45 Beacon Street, Boston, MA 02108-3693. (http://www.ametsoc.org/AMS/)

ASCE: American Society of Civil Engineers, 1801 Alexander Bell Drive, Reston, VA 20191-4400. (http://www.asce.org)

ASOS: Automated Surface Observing Station.

AWIPS: Advanced Weather Interactive Processing System.

BAMS: *Bulletin of the American Meteorological Society*, 45 Beacon Street, Boston, MA 02108-3693.

Broadcast seeding: The release of seeding agent, either from aircraft or from the ground, in conditions thought favorable for the development of treatable convective storms, but either before such storms have developed or at some distance from the storms (i.e., not within the storm or in its immediate proximity). Compare *direct targeting*. Broadcast seeding also is a routine practice to increase precipitation within winter clouds in mountainous areas.

Burn-in-place flare: A pyrotechnic device burned in a fixed position, such as the trailing edge of an aircraft wing. Compare *ejectable flare*.

CCN: Cloud Condensation Nuclei. Tiny solid and/or liquid (not predominantly pure water) *nuclei* upon which water vapor first condenses as the relative humidity approaches 100%. These *nuclei* are the seeds for cloud droplets.

Cell: A convective element (cloud), which, in its life cycle, develops, matures, and dissipates, usually in about 20 to 30 minutes. A cell in radar usage is a local maximum in radar reflectivity that undergoes a life cycle of growth.

Certified Weather Modification Manager: Certification of weather modification project managerial experience and skills granted by the WMA.

Certified Weather Modification Operator: Certification of weather modification project operational experience and skills granted by the WMA.

Cloud condensate: Liquid and ice water present in clouds.

Cloud droplets: Liquid water droplets, which are too small to precipitate since they typically range from a micron to tens of microns in diameter. Such droplets suspended in the atmosphere with other droplets form a cloud. Note: human hair has an average thickness of about 65 microns.

Cloud model: Physical description of cloud processes programmed into a computer to simulate cloud development and evolution. Useful in understanding the relative importance of the many factors that influence cloud development, and the only way in which exactly the same cloud can be both seeded and unseeded.

Cloud type: Cloud conditions with distinctive physical characteristics (cloud depth, temperature, wind, stability, etc.) such that different types may have different responses to seeding.

CO_2: See *dry ice*.

Coalescence: In cloud physics, the merging of two colliding water drops into a single drop. Coalescence between colliding drops is affected by the impact energy, which tends to increase with the higher fall velocities of larger drops.

Colloid: A mixture composed of two phases of matter, the dispersed phase (or colloid) and the continuous phase (or the dispersion medium). Colloids do not settle. A system of liquid and/or solid nuclei (i.e., have diameters less than 1 micron) colloidally dispersed in air is called an aerosol. In the case of fog, the dispersed phase is liquid and the continuous phase is air, but since its liquid components are generally greater than 1 micron in diameter it is not technically a colloidal system.

Conceptual model: A theoretical model of precipitation, hail, and/or fog development and the seeding methods to enhance or mitigate that development, based on current knowledge and scientific concepts. See also *cloud model.*

Contamination: The inadvertent distribution of seeding agent into areas that, according to project design, were not to have been seeded.

Continental air mass: A large body of air that forms over a continent, and consequently has the basic continental characteristic of relatively low water vapor content. See *Air mass.*

Control area: Areas where cloud seeding operations do not take place, preferably similar in character and near to the *target area.* The behavior of storms over the control area is compared to those treated over the target area to assess differences and thus measure project effectiveness. See also *target area* and *seeding area.*

CONUS: Continental United States.

Convective cloud: A cloud characterized by organized, fluid motion, including both upward and downward motions. This term generally is interchangeable with cumulus cloud.

CRAFT: Collaborative Radar Acquisition Field Test.

CRBPP: Colorado River Basin Pilot Project.

Crossover design: A project that employs areas that alternate between target and control. This crossover reduces the possibility of geographically induced bias in the evaluation.

Cumulus cloud: A principal cloud type in the form of individual, detached elements that are generally dense, and have characteristically sharp, non-fibrous outlines. These elements develop vertically, appearing as rising mounds, domes, or towers, the upper parts of which often resemble a cauliflower.

Direct targeting: The placement of seeding agents directly into the target cloud mass, either by release during penetration by aircraft, rocket, or artillery, or from aircraft flying directly below cloud base in updraft. Compare *broadcast seeding.*

Drizzle drop: A drop of water with diameter between 0.2 mm and 0.5 mm, which usually (but not always) falls from stratus or stratocumulus clouds. Drizzle is sometimes popularly called mist.

Droplet spectrum: The numbers and sizes of the droplets within the cloud volume of interest.

Dry ice: Frozen or solidified carbon dioxide (CO_2). Dry ice pellets have an equilibrium surface temperature of −78°C (at ambient pressure), and an operational temperature range colder than −2°C. Dry ice is used fairly often, especially in applications where environmental concerns about silver doses are heightened or the temperature is between 0°C and −5°C.

Dynamic seeding: The treatment of clouds with the intent of using the latent heat produced by additional freezing and perhaps in some cases by condensation or deposition to invigorate cloud development.

Ejectable flares: Pyrotechnic devices that are ignited and released (ejected) from aircraft. Compare *burn-in-place flare.*

Entrainment: The mixing of environmental air into a preexisting organized air current so that the environmental air becomes part of the current; for example, the entrainment of air into cumulus clouds. Entrainment of air into clouds, especially cumulus, is said to be inhomogeneous when the timescale for mixing of environmental air is very much greater than the timescale for droplet evaporation. Entrainment deepens the mixed layer in the absence of advection effects.

Environmental Impact Statement (EIS): A document prepared by a governmental agency proposing a project that states the environmental impacts which may affect the quality of the human environment.

Evaporation-mixing: Denotes a process in which condensation occurs following the mixture of two different air parcels. The mixture is commonly driven by the diffusion of the vapor from the warmer mixture into the other. Also associated with fog formation.

EWRI: Environmental and Water Resources Institute, American Society of Civil Engineers, Reston, VA. (http://www.ewrinstitute.org)

FAA: Federal Aviation Administration. The governmental entity that regulates aircraft operations, safety, and use of airways in the United States. Analogous entities exist in most other nations. (http://www.faa.gov)

FACE: Florida Area Cumulus Experiment.

Fog droplets: See *cloud droplets.*

Geostationary satellite: A satellite in a west-east orbit at an altitude of 35,786 km above the equator. At this altitude, its orbit matches the earth's rotation such that the satellite can be maintained over the same ground location. A geostationary satellite orbit is not necessarily the same as a geosynchronous orbit.

Geosynchronous satellite: A satellite in an equatorial or near equatorial orbit that orbits at the same angular velocity as the earth, making one revolution in 24 hours. A *geosynchronous* orbit is not necessarily the same as a *geostationary* orbit.

Glaciogenic: Causing the formation of ice.

Glaciogenic seeding: Treatment of clouds with materials intended to increase and/or initiate the formation of ice crystals.

GOES: Geostationary Operational Environmental Satellite. These are the latest NOAA weather satellites currently operational over the continental United States.

GPS: Global Positioning System. A global, satellite-based navigation positioning system that provides consistently accurate positions, based on a constellation of 24 low earth-orbiting satellites with very accurate clocks and the computational resources to triangulate the positions near the earth's surface. The system was developed by the U.S. Department of Defense for one satellite to determine position with an accuracy of 30 to 100 m, and accuracy within millimeters of a known reference position if two satellites are used and integration times are sufficiently long.

Graupel: White, opaque, heavily rimed snow particles that are about 2 to 5 mm in diameter. Also known as snow pellets, they form in convective clouds when supercooled water droplets freeze to an ice particle on impact. Graupel are sometimes distinguishable by shape, such as conical, raspberry, and lump (irregular) graupel.

Heterogeneous nucleation: The phase change of a substance to a more condensed state (i.e., a lower thermodynamic energy state) initiated by nuclei with different physical and chemical properties than the substance. For example, the nucleation of ice crystals from supercooled water vapor using silver iodide as the nuclei. See *nucleation.*

HIPLEX: HIgh PLains EXperiment. Part of the Bureau of Reclamation's Project Skywater.

Homogeneous nucleation: The phase change of a substance to a more condensed state (i.e., a lower thermodynamic energy state) initiated by nuclei with the same physical and chemical properties as the substance. That is, the nucleation system only contains one component. For example, the nucleation of ice crystals from supercooled water vapor. See *nucleation.*

Hydrometeor: Any product of condensation or deposition of atmospheric water vapor, whether formed in the free atmosphere, at the earth's surface, including being derived from a wind-blown water surface. Thus for example, a hydrometeor is cloud drops or ice particles of any size and shape, either suspended in the air or precipitating.

Hygroscopic: Pertaining to a marked ability to accelerate the condensation of water vapor. The ability of nuclei to absorb vapor but at a low enough rate that it does not completely dissolve under most conditions. This term is principally applied, in meteorology, to those condensation nuclei composed of salts that yield aqueous solutions of a very low equilibrium vapor pressure compared with that of pure water at the same air temperature.

Hygroscopic seeding: Treatment of clouds with hygroscopic materials that encourage the formation of larger droplets, changing the cloud droplet spectrum in such a way as to enhance development of precipitation through coalescence.

Ice Forming Nucleus: See *IFN.*

Ice nucleus: Any particle that serves as a nucleus for the formation of ice crystals in the atmosphere. The subset of atmospheric particles (essentially all airborne matter) upon which ice crystals will form. These nuclei are typically water insoluble particles, and may be classified as hydrophobic condensation nuclei. These are sometimes abbreviated IN or IFN, which stand for ice nuclei or ice forming nuclei, respectively.

Ice process: The process by which cloud particles grow large enough to fall out as ice-phase precipitation. This often occurs where there is coexistence of ice and supercooled water droplets. The ice particles can grow rapidly at the expense of the supercooled water droplets.

ICPMS: Inductively Coupled Plasma-Mass Spectrometer; a relatively new technique to determine the concentration of trace elements in solution.

IFN: Ice forming nucleus. See *ice nucleus.*

Immunity: In a legal liability action against a government, a defense based on the concept that the government cannot legally be sued.

IN: See *ice nucleus.*

Inadvertent weather modification: The unintentional modification of the weather through some aspect of man's activities, such as the production of cloud nuclei or ice nuclei from various industrial-manufacturing processes.

Indemnification: Payment to a person or agency of an amount of money equal to any loss it may have incurred including any legal liability payment made.

In-situ measurements: In-situ measurements refer to the gathering of information from within or on a media. That is, they are made in the actual location or environment of the object or entity measured. Such measurements are still the most common type of measurements, although *remote sensing measurements* are gaining on them.

Instrument flight rules (IFR): The FAA regulations pertaining to flight at altitudes of 5.5 km (18,000 feet) above mean sea level or higher over U.S. airspace, or in any meteorological conditions necessitating the use of aircraft instrumentation for safe navigation.

JWM: Journal of Weather Modification. The official journal of the Weather Modification Association. See *WMA.*

KCl: See *potassium chloride.*

Latent heat: The heat released or absorbed per unit mass by a system in a reversible, isobaric-isothermal change of phase. Simply, the heat released into the surrounding air or absorbed from the surrounding air when water changes its physical state. For example, the heat released when water vapor condenses is called latent heat of condensation; the heat released when a liquid water droplet freezes is called the latent heat of fusion.

License: Document issued by a government agency to an individual authorizing the holder to practice a profession.

Lidar: Light detection and ranging. An instrument combining a pulsed laser transmitter and optical receiver (usually a telescope) with an electronic signal processing unit used for the detection and ranging of various targets within the atmosphere, such as atmospheric particles, analogous to the principles of operation of microwave radar.

LORAN: LOng RAnge Navigation.

Maritime air mass: A large body of air that forms over an ocean. See *Air mass.*

MOA: Military operations area.

MSDS: Material Safety Data Sheets.

NaCl: See *sodium chloride.*

NAIWMC: North American Interstate Weather Modification Council, 20 Moore Lane, Reno, NV, 89509. (www.naiwmc.org)

NASS: National Agricultural Statistics Service.

NCAR: National Center for Atmospheric Research, P.O. Box 3000, Boulder, CO, 80305. (www.ncar.ucar.edu/ucar/index.html)

NDTP: North Dakota Thunderstorm Project.

Necessity: In a legal liability action a defense against liability based on the privilege to undertake activities to prevent an imminent public disaster.

Negligence: Careless conduct falling below the standard of reasonable prudent care.

NEXRAD: See *WSR-88D.*

NH_4I: Chemical formula for ammonium iodide.

NOAA: National Oceanic & Atmospheric Administration, U.S. Department of Commerce. The parent organization of the U.S. National Weather Service, and the federal agency to which all U.S. Weather Modification activities must be reported. (http://www.noaa.gov)

Nowcasting: Very short-term forecasting, from the present to about 30 minutes.

Nucleation: Any process of initiating a phase change of a substance to a more condensed state (or a lower thermodynamic energy state). Some examples include, the process of initiating the ice phase in a supercooled liquid water drop, and the process of initiating the liquid phase from its vapor. The nucleation may be *homogeneous nucleation* or the more common *heterogeneous nucleation.*

Nuclei: Airborne particles, droplets, or vapor/gas molecules upon or through which a phase change may occur. A basis for future development and growth; a kernel. See *CCN, ice-forming nucleus, ice nucleus.*

Nuisance: Conduct that significantly harms the right to use and enjoy properly and which on the balance is less desirable than the benefits obtained from it.

NWS: National Weather Service. See *NOAA.*

Operational area: Contains the target area and any surrounding upwind areas employed for treatment in order that the effect from seeding covers the target area, as well as for evaluating the project (excluding the control area as long as these areas have not had seeding conducted in them).

Operational cloud seeding project: A cloud seeding project conducted for a specific purpose, such as optimizing the production of precipitation within a target area.

Opportunity recognition: The identification of those clouds or cloud systems suitable for seeding according to the conceptual model.

Orographic: Relating to mountains or mountain effects. Most often refers to the influences of mountains or mountain ranges on wind field, but also to describe the effects on other meteorological quantities such as air temperature, humidity, or precipitation distribution.

Overseeding: A condition that results from the application of too much glaciogenic seeding agent, in which case too many small ice crystals may form, none of which are large enough to precipitate or aggregate.

Permit: A document issued by a government agency to an individual, group, or entity authorizing a holder to carry out some activity as specified therein.

Placebo: Treatment with an inert substance, without the knowledge of those applying the treatment. In a randomized cloud seeding program, clouds are treated with real seeding agents or a placebo, which might be only an audible event such as a recorded "bang" that sounds like a flare firing, or a flare containing sand.

Potassium chloride (KCl): A simple salt often used as a primary ingredient in hygroscopic cloud seeding pyrotechnics.

PPI: Plan position indicator.

PPT: Parts per trillion.

Precipitation efficiency: The efficiency at which condensed water within a cloud is transformed into hydrometeors that can fall out and reach the ground as precipitation.

Project area: The area covered by all the equipment used for operations within all other areas used in evaluation or assessment by the contractors and/or sponsors.

Pyrotechnic: "Fireworks," i.e., in this manual, a flare that burns to produce either AgI or hygroscopic nuclei.

Radiation: The transfer of heat energy without the involvement of a physical substance, such as air molecules. For example, the heat from a stove burner is an example of radiation because the heat would flow away from this burner regardless of whether we had air molecules, no air molecules (as in space), a body of water, water and a piece of metal, or a piece of metal between the burner and the receptor. Radiative cooling of an atmospheric layer could lead to additional cloud formation and possibly precipitation given specific circumstances.

Radiosonde (or rawinsonde): An expendable meteorological instrument package that senses and transmits temperature, dew point temperature, and pressure at various altitudes during its ascent into the atmosphere. These quantities can be converted into relative humidity. A radiosonde is carried aloft by weather balloons twice daily from many sites around the world, and may also be employed by projects to bolster forecasting or research efforts. If their position is tracked in time then they will also reveal the vertical wind field profile. The radiosonde that is tracked in time to also obtain wind field is known as a rawinsonde.

Raindrop: A drop of water of diameter greater than 0.5 mm. See *drizzle drop.*

RAP: Research Applications Program.

RASS: Radio acoustic sounding systems.

Remote sensing: A method of obtaining information about properties of an object or environment without coming into physical contact with that object or environment. Compare *in situ measurement.*

Remote sensing devices: Refers to any sensor(s) on satellite, airborne, or ground-based platforms that either directly (active) or indirectly (passive) infer information about a property of an environment without being in or attached to that environment. Examples of ground-based remote sensing devices include lidars, radars, radiometers, acoustic sounders, and sonic anemometers. Satellites have radiometers sensitive to one or more spectral regions of the electromagnetic spectrum on board a satellite. Some of the latest satellite platforms include radar, and will include a lidar in the near future. These devices can also be mounted on aircraft. Contrast with *in-situ measurements.*

Remote sensing measurements: Refers to the measurements from *remote sensing devices.*

Research cloud seeding projects: Cloud seeding projects organized primarily to acquire additional knowledge on how seedable the clouds in a given location might be; the precipitation processes that occur naturally and how they may be altered by seeding; the testing of different seeding modes, etc. Assessment of seeding effects is a primary interest. These projects are typically sponsored by government agencies because of their cost.

Response time: The time that elapses from identification of a seeding opportunity until the release of seeding agent actually begins. The response time, with respect to measuring devices, refers to the time interval necessary for the measuring device exposed to a change in an atmospheric property to reach the fraction $[1 - (1/e)]$ or 63.2% of the total environmental change in that atmospheric property that it would exhibit after an infinitely long time.

Revocation: Permanent cancellation of a license or permit.

RHI: Range height indicator.

SAD: Seasonal affective disorder.

SCPP: Sierra Cooperative Pilot Project.

Seeding agents: Agents dispensed by any means in or near a cloud volume which are intended to modify (seed) the cloud characteristics.

Seeding area: The area over which seeding operations are permitted. This includes the *target area* and additional area outside the target area to allow for seeding upwind for intended effect in the target. See also *control area.*

Seeding criteria: A set of conditions established for a cloud-seeding project that are designed to optimize the augmentation of precipitation. Typical indices used are cloud temperatures, wind flow, atmospheric stability, and water content.

Seeding hypothesis: A statement of expectations, which states certain assumptions and predicts an outcome in terms of a seeding effect, given specifics on a seeding project, such as seeding mode and a project location and duration.

Silver iodide (AgI): A common *glaciogenic* seeding agent. Its chemical symbol is AgI. See *ice nucleus.*

SLW: Supercooled liquid water. Supercooled means temperatures below 0°C.

SODAR: Sonic detection and ranging. These systems are used to remotely measure the vertical turbulence structure and wind profile of the lower layer of the atmosphere.

Sodium chloride (NaCl): The chemical composition of common table salt. Salt powder is being used for hygroscopic seeding, because of its hygroscopic properties.

Static cloud seeding: Cloud seeding to alter the precipitation by changing the efficiency with which existing cloud water is converted into precipitation-size particles.

Stratiform clouds: Descriptive of clouds with limited vertical development and extensive horizontal development, as contrasted to the vertically developed cumuliform cloud types.

Stratus clouds: These clouds are uniformly stratified and nearly always have a uniform base. They are normally precipitation free, but occasionally drizzle, or light mist will fall out of stratus clouds. There are weather reports of light to moderate continuous rain from stratus clouds that form near an ocean.

Supercooled water: Water that remains in the liquid state despite air temperatures colder than 0°C. Some pure water droplets may exist in a supercooled state to temperatures well below −40°.

Suspension: Temporary cancellation of a license or permit. A temporary halt of seeding operations to avoid undesirable results. See *suspension criteria.*

Suspension criteria: Criteria developed for a specific cloud seeding project to avoid seeding during undesirable periods. Examples include excess snowpack accumulation, excess rainfall forecasts, and National Weather Service forecasts of severe weather.

Target area: The region for which cloud seeding operations are targeted.

Targeting: The releasing of artificial cloud seeding material in a manner that allows adequate dispersion of the material, interception of supercooled liquid water droplets, growth of ice crystals, and fallout of augmented precipitation in a specified target.

Terminal velocity: The particular falling speed for any given object (i.e., a raindrop) moving through a fluid of specified physical properties (cloudy air), at which the drag forces and buoyant forces exerted by the fluid on the object (i.e., a raindrop) just equal the gravitational force acting on the object (i.e., a raindrop).

Thermal: A relatively small-scale, rising current of air produced when the atmosphere is heated enough locally by the earth's surface to produce absolute instability in the lowest layers.

TITAN: Thunderstorm Identification, Tracking, Analysis, and Nowcasting. Software for the display and analysis of weather radar data; widely used in operational convective cloud seeding programs.

Trespass: Intended intrusion upon the land of another person.

Turbulence: Irregular atmospheric motion, especially when characterized by upward and downward currents.

USAF: United States Air Force.

UV: Ultraviolet electromagnetic radiation of shorter wavelength than visible radiation, but longer than X-rays. Ultraviolet radiation (light) is a component of normal solar radiation. Foam hail pads will degrade with prolonged exposure to UV radiation, and are either covered with foil or painted.

Wilderness area: Area formally designated by federal statute as one which must be kept in wilderness status.

Wind field: The three-dimensional space (i.e., vertical and horizontal) and temporal values of wind speed and wind direction over a surface on the earth, which is usually continuous except at surfaces or curves.

Wing-tip generator: Cloud seeding generators mounted at or near the tips of aircraft wings.

WMA: Weather Modification Association, P.O. Box 26926, Fresno, CA 93729-6926. (http://www.weathermodification.org)

WMO: World Meteorological Organization, 7 bis Avenue de la Paix, CP 2300-1211, Geneva 2-Switzerland.
(http://www.wmo.ch/index-en.html)

WSR-88D: The 1988 vintage Doppler weather radar network by NEXRAD; deployed in the United States by the National Weather Service during the 1990s.

INDEX